Introduction to Financial Forecasting in Investment Analysis

John B. Guerard, Jr.

Introduction to Financial Forecasting in Investment Analysis

Springer

John B. Guerard, Jr.
McKinley Capital Management, LLC
Anchorage, AK, USA

ISBN 978-1-4614-5238-6 ISBN 978-1-4614-5239-3 (eBook)
DOI 10.1007/978-1-4614-5239-3
Springer New York Heidelberg Dordrecht London

Library of Congress Control Number: 2012952930

Printed on acid-free paper

Springer is part of Springer Science+Business Media (www.springer.com)

Preface

An Introduction to Financial Forecasting in Investment Analysis

The objective of this proposed text is a 250 page introductory financial forecasting text that exposes the reader to applications of financial forecasting and the use of financial forecasts in making business decisions. The primary forecasts examined in this text are earnings per shares (eps). This text will make extensive use of I/B/E/S data, both historic income statement and balance sheet data and analysts' forecasts of eps. We calculate financial ratios that are useful in creating portfolios that have generated statistically significant excess returns in the world of business. The intended audience is investment students in universities and investment professionals who are not familiar with many applications of financial forecasting. This text is a data-oriented text on financial forecasting, understanding financial data, and creating efficient portfolios. Many regression and time series examples use E-Views, OxMetrics, Scientific Computing Associates (SCA), and SAS software.

The first chapter is an introduction to financial forecasting. We tell the reader why one needs to forecast. We introduce the reader to the moving average and exponential smoothing models to serve as forecasting benchmarks.

The second chapter introduces the reader to the regression analysis and forecasting. In the third chapter, we use regression analysis to examine the forecasting effectiveness of the composite index of leading economic indicators, LEI. Economists have constructed leading economic indicator series to serve as a business barometer of the changing US economy since the time of Wesley C. Mitchell (1913). The purpose of this study is to examine the time series forecasts of composite economic indexes, produced by The Conference Board (TCB) and test the hypothesis that the leading indicators are useful as an input to a time series model to forecast real output in the USA. Economic indicators are descriptive and anticipatory time-series data are used to analyze and forecast changing business conditions. Cyclical indicators are comprehensive series that are systemically related to the business cycle.

The third chapter introduces the reader to the forecasting process and illustrates exponential smoothing and (Box–Jenkins) time series model estimations and forecasts using the US Real Gross Domestic Product (GDP). The chapter is a "hands-on" exercise in model estimating and forecasting. In this chapter, we examine the forecasting effectiveness of the composite index of leading economic indicators, LEI. The leading indicators can be an input to a transfer function model of real Gross Domestic Product, GDP. The transfer function model forecasts are compared to several naïve models in terms of testing which model produces the most accurate forecast of real GDP. No-change forecasts of real GDP and random walk with drift models may be useful as a forecasting benchmark (Mincer and Zarnowitz 1969; Granger and Newbold 1977).

The fourth chapter addresses the issue of composite forecasting using equally weighted and regression-weighted models. We discuss the use of GDP forecasts. We analyze a model of United States equity returns, the USER Model, to address issues of outliers and multicollinearity. The USER Model combines Graham & Dodd variables, such as earnings, book value, cash flow, and sales with analysts' revisions, breadth, and yields and price momentum to rank US equities and identify undervalued securities. Expected returns modeling has been analyzed with a regression model in which security returns are functions of fundamental stock data, such as earnings, book value, cash flow, and sales, relative to stock prices, and forecast earnings per share (Fama and French 1992, 1995; Bloch et al 1993; Haugen and Baker 2010; Stone and Guerard 2010).

In Chapter 5, we expand upon the time series models of Chap. 2 and introduce the reader to multiple time series model and Granger causality testing as in the Ashley, Granger, and Schmalensee (1980) and Chen and Lee (1990) tests. We illustrate causality testing with mergers, stock prices, and LEI data in the USA in the postwar period.

In Chapter 6, we examine analysts' forecasts in portfolio construction and management. We use the Barra risk optimization analysis system, the standard portfolio risk model in industry, to create efficient portfolios. The Barra Aegis system produces statistically significant asset selection using the USER Model for the 1980–2009 period.

In Chapter 7, we show how US, Non-US, and Global portfolio returns can be enhanced by use of eps forecasts and revisions. We use the Sungard APT and Axioma systems to create efficient portfolios using principal components-based risk models. McKinley Capital Management hosted a research seminar in Anchorage in July 2011. The APT and Axioma results presented in Chapter 7 extend portfolio construction applications presented at the McKinley conference and published in the Spring 2012 *Journal of Investing* special edition on Quantitative Risk Models.

We illustrate global market timing and tactical asset management in Chapter 8. The ability to forecast market shifts allows the manager to increase his or her risk acceptance and enhance the risk-return tradeoff.

We summarize our processes, tests, and results in Chapter 9. We produce conclusions that are relevant to the individual investor and portfolio manager.

The author acknowledges the support of his wife of 30-plus years, Julie, and our three children, Richard, Katherine, and Stephanie. The author gratefully acknowledges the comments and suggestions of several gentlemen who each read several chapters of this monograph. Professors Derek Bunn, of the London Business School, Martin Gruber, of New York University, Dimitrios Thomakos, of the University of Peloponnese (Greece), and Jose Menchero of MSCI Barra. Special thanks to Anureef Saxena, Robert Stubbs, and Dewitt Miller for suggestions and edits of Chapters 7 and 8. Robert (Rob) Gillam, the Chief Investment Officer of McKinley Capital Management, is acknowledged for his support of APT, which he brought into our firm seven years ago when he hired me, and Axioma, which we purchased last year. Any errors remaining are the responsibility of the author.

Anchorage, AK, USA John B. Guerard, Jr.

Contents

Chapter 1
Forecasting: Its Purpose and Accuracy

The purpose of this monograph is to concisely convey forecasting techniques to applied investment analysis. People forecast when they make an estimate as to the future value of a time series. That is, if I observe that IBM has a stock price of \$205.48, as of March 23, 2012, and earned an earnings per share (eps) of \$13.06 for fiscal year 2011, then I might wonder at what price IBM would trade for on December 31, 2012, if it achieved the \$14.85 eps that 21 analysts, on average, expect it to earn in 2012 (source: MSN, Money, March 23, 2012, 1:30 p.m., AST). The low estimate is \$14.18 and the high estimate is \$15.28. Ten stock analysts currently recommend IBM as a "Strong Buy," one as a "Moderate Buy," and ten analysts recommend "Hold." Moreover, if IBM achieves its forecasted \$16.36 eps average estimate for December 2013, when could be its stock price and should an investor purchase the stock? One sees several possible outcomes; can IBM achieve its forecasted eps figure? How accurate are the analysts' forecasts? Second, should an investor purchase the stock on the basis of an earnings forecast? Is there a relationship between eps forecasts and stock prices? How accurate is it necessary for analysts to be for investors to make excess returns (stock market profits) trading on the forecasts?

Granger (1980a, b) differentiated between an event outcome such as to forecast IBM eps (at a future date), event time, such as whether the US economy will completely recover from the 2008 to 2009 recession and IBM realize its forecasted eps, and time series forecasts, generating the forecasts and confidence intervals of IBM earnings at future dates. In this monograph, we concentrate on using eps forecasts for IBM and approximately 16,000 other firms in stock selection modeling and portfolio management and construction strategies to generate portfolio returns that outperform the portfolio manager benchmark. To access the effectiveness of producing and using forecasts, it is necessary to establish forecast benchmarks, measures of forecast accuracy, and methods to test for effective forecast implementation.

One can establish several reasonable benchmarks for forecasting. First, the use of a no-change model, in which last period's value is used as the forecast for the current period forecast, has a long and well-recognized history [Theil (1961 and 1966)

J.B. Guerard, Jr., *Introduction to Financial Forecasting in Investment Analysis*,
DOI 10.1007/978-1-4614-5239-3_1,

and Mincer and Zarnowitz (1969)]. Second, one can establish several criteria for forecast accuracy. The forecast error, e_t, is equal to the actual value, A_t, less the forecasted value, F_t. One can seek to produce and use forecasts that have the lowest errors on the following measurements:

$$\text{Mean Error} = \frac{\Sigma_{t=1}^{T} e_T}{T};$$

$$\text{MAPE} = \text{Mean Absolute Percentage Error} = \frac{\Sigma_{t=1}^{T} |e_t|}{T};$$

and

$$\text{Mean Squared Forecast Error} = \text{MSFE} = \sum_{t=1}^{T} e_t^2.$$

There are obviously advantages and disadvantages to these measures. First, in the mean error calculation, small positive and negative values may "cancel" out implying that the forecasts are "perfect." Makridakis and Hibon (2000) remind us that the mean error is only useful in determining whether the forecaster over-forecasts, producing positive forecast errors; that is, the forecaster has a positive forecast bias. The MAPE is the most commonly used forecast error efficiency criteria [Makridakis, Anderson, Carbone, Fildes, Hibon, Newton, Parzen, and Winkler (1984)]. The MAPE recognizes the need of the forecast to be as close as possible to the realized value. Thus, the sign of the forecast error, whether positive or negative, is not the primary concern. Finally, the mean squared forecast error is assuming a quadratic loss function, that is, a large positive forecas error is not preferred to a large negative forecast error. In this monograph, we examine the implications of the three primary measures of forecast accuracy. We are concerned with two types of forecasts: the economy (the United States and the World, particularly the Euro zone) and analysts forecasts of corporate eps. Why? We believe, and will demonstrate, that a reasonable economic forecast of the direction of the economic strength is significant in allowing an asset manager or an investor to participate in economic growth. Second, we find that firms achieving the highest growth in eps generate the highest stock holder returns during the 1980–2009 period; moreover, we will demonstrate that the securities that achieve the highest eps growth and hence returns are not those forecast to have the highest eps, but are not that have the highest eps forecast revisions and that it is equally important for analysts to agree on the eps revisions. That is, the larger the number of analysts that raise their respective eps forecasts, the highest will be stockholder returns.

The purpose of this monograph is to introduce the reader to a variety of financial techniques and tools to produce forecasts, test for forecasting accuracy, and demonstrate the effectiveness of financial forecasts in stock selection, portfolio construction and management, and portfolio attribution. We believe that financial markets are very near to being efficient, but statistically significant excess returns can be earned.

Let us discuss several aspects to forecast accuracy: forecast rationality, turning point analysis, and absolute and relative accuracy.

Forecast Rationality

One of the most important aspects of forecast accuracy is forecast rationality. Clements and Hendry (1998) discuss rationality in several levels. "Weak" rationality is associated with the concept of biasedness. A test of unbiasedness is generally written in the form

$$A_t = \alpha + \beta P_t + \varepsilon_t, \tag{1.1}$$

where

A_t, actual value at time t;

P_t, predicted value (forecast) at time t;

ε_t, error term at time t.

In (1.1), we have only assumed a one-step-ahead forecast horizon. One can replace t with $t + k$ to address the issues of $k =$ Period ahead periods. Unbiasedness is defined in (1.1) with the null hypothesis that $\alpha = 0$ and $\beta = 1$. The requirement for unbiasedness is that $E(\varepsilon_t) = 0$. In expectational terms

$$E[A_t] = \alpha + \beta E[P_t]. \tag{1.2}$$

One expects $\beta = 1$ and $\alpha = 0$, a sufficient, but not necessary condition for unbiasedness. "Strong" rationality or efficiency requires that the forecast errors are uncorrelated with other data or information available at the time of the forecast, Clements and Hendry (1998).

Much of forecasting analysis, measurement, and relative accuracy was developed in Theil (1961) and Mincer and Zarnowitz (1969). Theil discussed several aspects of the quality of forecasts. Theil (p. 29) discussed the issue of turning points, or one-sided movements, correctly. Theil produced a two-by-two dichotomy of turning point forecasting. The Theil turning point analysis is well worth reviewing. A turning point is correctly predicted; that is, a turning point is predicted and an actual turning point occurs (referred as "i"). In a second case, a turning point is predicted, but does not occur ("ii"). In the third case, a turning point actually occurs, but was not predicted ("iii"); the turning point is incorrectly predicted. In the fourth and final case, a turning point is not predicted and not recorded. Thus, "i" and "iv" are regarded as forecast successes and "ii" and "iii" are regarded as forecast failures. The Theil turning point table is written as

Actual turning points		Predicted turning points
	Turning point	No turning point
Turning point	i	iii
No turning point	ii	iv

The Theil turning point failure measures:

$$\phi_1 = \frac{iii}{i+ii}; \quad \phi_2 = \frac{iii}{i+iii}.$$

Small values of ϕ_1 and ϕ_2 indicate successful turning point forecasting. The turning point errors are often expressed graphically, where

Chart 1.

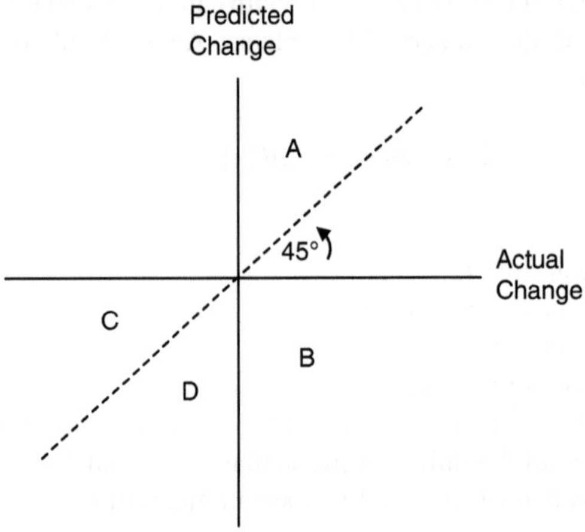

Regions *A* and *D* represent overestimates of changes whereas regions *B* and *C* represent underestimates of changes. The 45° line represents the line of perfect forecasts. Elton, Gruber, Brown, and Goetzmann (2009) make extensive use of the Theil graphical chart in their analysis of analysts' forecasts of eps.

A line of perfect forecasting is shown in Chart 2, where $U = 0$.

Chart 2.

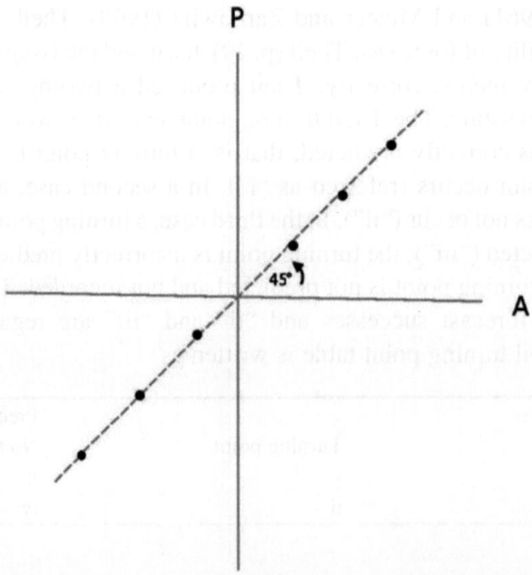

A line of maximum inequality is shown in Chart 3 where $U = 1$.

Chart 3.

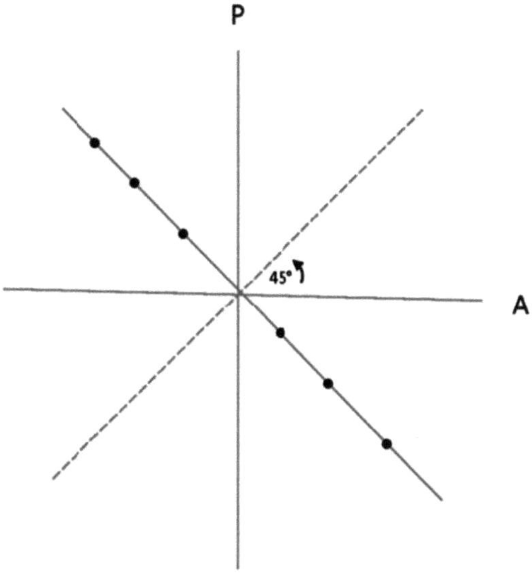

The forecasters in Chart 3 are very bad (the worst possible). Intermediate grades of forecasting are shown in Chart 4 and Chart 5 where the respective μ are small and large, respectively.

Chart 4.

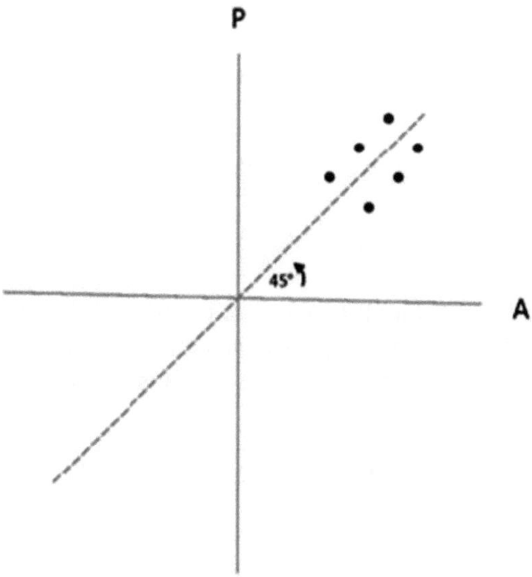

Chart 5.

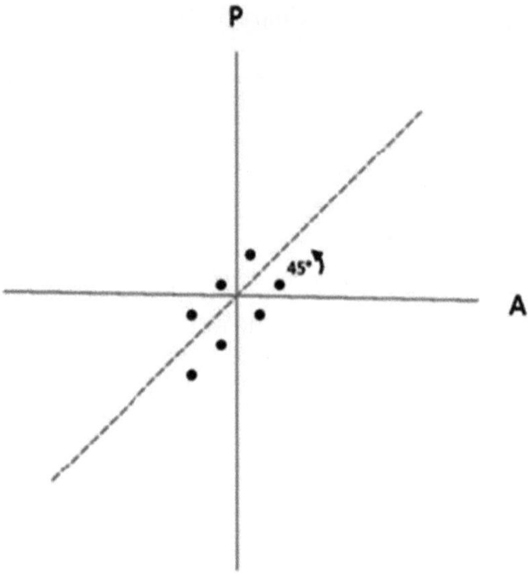

Theil (1961, p. 30) analyzed the relationship between predicted and actual values of individual i.

$$\mathrm{P}_i = \alpha + \beta A_i, \quad \beta > 0. \tag{1.3}$$

Perfect forecasting requires that $\alpha = 0$ and $\beta = 1$. An alternative representation of (1.3) can be represented by the now familiar inequality coefficient, now known as Theil's U, or Theil Inequality coefficient, TIC.

$$\mu = \frac{\sqrt{\frac{1}{T}\Sigma(P_i - A_i)^2}}{\sqrt{\frac{1}{T}\Sigma P_i^2} + \sqrt{\frac{1}{T}\Sigma A_i^2}}. \tag{1.4}$$

If $U = 0$, then $P_i = A_i$ for all I, and there is perfect forecasting. If $U = 1$, then the TIC reaches its "maximum in equality" and this represents very bad forecasting. Theil broke down the numerator of μ into sources or proportions of inequality.

$$\frac{1}{T}\Sigma(P_i - A_i)^2 = (\bar{P} - \bar{A})^2 + (S_P - S_A)^2 + 2(1 - r)S_P S_A, \tag{1.5}$$

where
$\bar{P}$ = mean of predicted values;

$\bar{A}$ = mean of actual values;
S_P = standard deviation of predicted values;
S_A = standard deviation of actual values;
and
r = correlation coefficient of predicted and actual values.
Let D represent the denominator of (1.4).

$$U_M = \frac{\bar{P} - \bar{A}}{D};$$

$$U_S = \frac{S_P - S_A}{D};$$

$$U_C = \frac{\sqrt{2(1-r)S_P S_A}}{D};$$

$$U_M^2 + U_S^2 + U_C^2 = U^2. \tag{1.6}$$

The term U_M is a measure of forecast bias. The term U_S represents the variance proportion and U_C represents the covariance proportion. U_M is bounded within plus and minus 1; that is, $U_M = 1$ indicates no variation of P and A or perfect correlation with slope of 1.

$$U^M = \frac{U^2 M}{U^2}; \quad U^S = \frac{U_S^2}{U^2} = U^C = \frac{U_C^2}{U^2}.$$

Theil refers to U^M, U^S, and U^C as partial coefficients of inequality due to unequal central tendency, unequal variation, and imperfect correlation, respectively.

$$U^M + U^S + U^C = 1. \tag{1.7}$$

Theil (1961, p. 39) decomposes (1.5) into

$$\frac{1}{T}\Sigma(P_i - A_i)^2 = (\bar{P} - \bar{A})^2 + (S_P - S_A)^2 + (1 - r^2)S_A^2. \tag{1.8}$$

If a forecast is unbiased, then $E(\bar{P}) = E(\bar{A})$ and, in the regression of

$$A_i = P_i + U_i,$$

where U_i = regression error term, the slope of A on P is $\frac{rS_A}{S_P}$. $U^2 = U_M^2 + U_R^2 + U_D^2$, where $U_R^2 = \left(\frac{S_P - rS_A}{D}\right)^2$;

$$U_D^2 = \left(\frac{\sqrt{(1-r^2)}S_A}{D}\right)^2.$$

U_R is inequality due to an incorrect regression slope and U_D is inequality due to nonzero regression error terms (disturbances).

$$U^R = \frac{U_R^2}{U^2} \quad \text{and} \quad U^D = \frac{U_D^2}{U^2}.$$

The U^R term is the regression proportion of inequality. The U^D term is the disturbance proportion of inequality.

$$U^M + U^R + U^D = 1.$$

The modern version of the TIC is written as the Theil U as

$$U = \sqrt{\frac{\Sigma_{t=1}^{T-1}\left(\frac{F_{t+1} - Y_t - Y_{t+1} + Y_t}{Y_t}\right)^2}{\Sigma_{t=1}^{T-1}\left(\frac{Y_{t+1} - Y_t}{Y_t}\right)^2}} \tag{1.9}$$

or

$$U = \sqrt{\frac{\Sigma_{t=1}^{T-1}(FPE_{t+1} - APE_{t+1})^2}{\Sigma_{t=1}^{T-1}(APE_{t+1})^2}},$$

where

$$FPE_{t+1} = \frac{F_{t+1} - Y_t}{Y_t} \text{ and}$$
$$APE_{t+1} = \frac{Y_{t+1} - Y_t}{Y_t},$$

where F = forecast and A = Actual values,
where FPE is the forecast relative change and APE is the actual relative change.

Absolute and Relative Forecast Accuracy

Mincer and Zarnowitz (1969) built upon the TIC analysis and discussed absolute and relative forecasting accuracy in a more intuitive manner.

Chart 6.

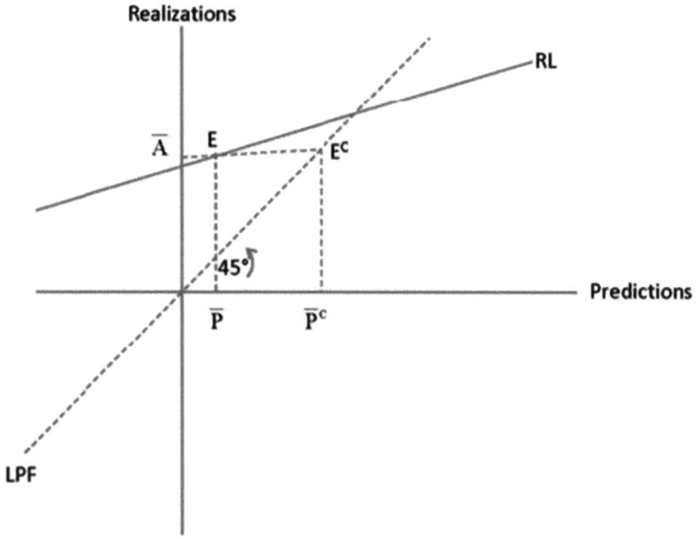

The line of perfect, LPF, is of course where $P = A$, as was the case with Theil. Mincer and Zarnowitz (1969) write the mean square error of forecast, M_P, as

$$M_P = E(A - P)^2, \quad (1.10)$$

where E denotes expected value. In the Mincer–Zarnowitz Prediction–Realization diagram, shown in Chart 6, the line $E - E^C$ denotes forecast bias. Thus, $E(A) - E(P) = E(U)$ denotes forecast bias.

Let us return for the actual–predicted value regression analysis:

$$A_t = P_t + u_t \quad (1.11)$$

which is estimated with an ordinary least squares regression of

$$A_t = \alpha + \beta P_t + v_t. \quad (1.12)$$

It is necessary for the forecast error, u_t, to be uncorrelated with forecast values, P_t, for the regression slope β to equal unity (1.0). The residual variance in the regression $\sigma^2(v)$ equals the variance of the forecast error $\sigma^2(u)$. Forecasts are efficient if $\sigma^2(u) = \sigma^2(v)$. If the forecast is unbiased, $\alpha = 0$, and $\sigma^2(v) = \sigma^2(u) = M_P$.

Mincer and Zarnowitz (1969) discuss economic forecasts in terms of predictions of changes (not absolute levels). The mean square error is

$$(A_t - A_{t-1}) - (P_t - A_{t-1}) = A_t - P_t = u_t. \quad (1.13)$$

The relevant Mincer–Zarnowitz regression slope is

$$\beta_\Delta = \frac{cov(A_t - A_{t-1}, P_t - A_{t-1})}{\sigma^2(P_t - A_{t-1})}.$$

If the level forecast is efficient, then $\beta = 1$ (cov $(u_t, P_t) = 0$). The $\beta_\Delta = 1$ and only if cov $(u,A_{t-1} = 0$. The extrapolative value of A_{t-1} must be incorporated into the forecasts. Underestimation of change occurs when the predicted change (P_t – A_{t-1}) is of the same size, but smaller size than the actual change (A_t – A_{t-1}).

$$E|P_t - A_{t-1}| < E|A_t - A_{t-1}| \tag{1.14}$$

or

$$E(P_t - A_{t-1})^2 < E(A_t - A_{t-1})^2.$$

$$[E(P_t) - E(A_{t-1})]^2 + \sigma^2(P_t - A_{t-1}) < [E(A_t) - E(A_{t-1})]^2 + \sigma^2(A_t - A_{t-1})^2. \tag{1.15}$$

Underestimation of changes occurs if

$$E(P_t) < E(A_t), \text{ when } A_t \text{ and } P_t > A_{t-1},$$
$$E(P_t) < E(A_t), \text{ when } A_t \text{ and } P_t < A_{t-1},$$

and or

$$\sigma^2(P_t - A_{t-1}) < \sigma^2(A_t - A_{t-1}). \tag{1.16}$$

In (1.16), when predictions of changes are efficient, $\beta_\Delta = 1$, then $\sigma^2(A_t - A_{t-1}) = \sigma^2(P_t - A_{t-1}) + \sigma^2(U_t)$.

Mincer and Zarnowitz (1969) decomposed the mean square error to create an index of forecasting quality, R_M. The index of forecasting quality is the ratio of the mean square error of forecast and the mean square error of extrapolation, the relative mean square error. If forecasts are "good" and are superior to extrapolated values, then $0 < R_M < 1$. If $R_M > 1$, then the forecast is inferior.

$$R_M = \frac{M_P}{M_X} = \frac{1 - \frac{U_X}{M_X}}{1 - \frac{UP}{M_P}} \times \frac{M_P^C}{M_X^C} = gRM^C. \tag{1.17}$$

If x is a best, unbiased, and efficient extrapolation then $M_X^C = M_X$ and $g = \frac{M_P}{M^CP} > 1$ and $RM^C \leq RM$. Mincer and Zarnowtiz found that autoregressive extrapolations were not optimal; however, $RM^C < RM$ in twelve of 18 cases. Mincer and Zarnowitz found that inefficiency was primarily due to bias.

Mincer and Zarnowitz put for the r a theory that if RM_C, the forecast is superior relative to an extrapolative forecast benchmark, then "useful autonomous information enhanced the forecast." Autoregressive extrapolations showed substantial improvement over naïve (average) models, and while not optimal, were thus more efficient. A small number of lags produced satisfactory extrapolative benchmarks.

The Mincer–Zarnowitz approach was important, not only because of its no-change benchmarks but (benchmark method of forecast) also because of its use of an extrapolative forecast which should incorporate the history of the series. Mincer and Zarnowitz concluded that the underestimation of changes reflects the conservative prediction of growth rates in series with upward trends.

Granger and Newbold (1977) addressed two aspects of Mincer and Zarnowitz. First in the Mincer and Zarnowitz forecast efficiency regression:

$$X_t = \alpha + \beta f_t + e_t. \tag{1.18}$$

A forecast is efficient if $\alpha = 0$ and $\beta = 1$. However, the forecast, f_t, must be uncorrelated with the error term, e_t. Granger and Newbold question this assumption in practical applications. Second, it is essential that the e_t, the error term be white noise-suboptimal forecasts (whether one-step-ahead or k-step-ahead) are not white noise. For a forecast to be optimal, the expected squared error must have zero mean and be uncorrelated with the predictor-series. Unless the error term series takes on the value "zero" with probability of one, the predictor series will have a smaller variance than the real series. Second, random walk series appear to give reasonable predictors of another independent random walk series. A random walk with drift forecast is the approximate form as a first-order exponential smoothing model shown in the appendix. We show the first-order and second-order exponential smoothing model, the linear, trend, and seasonal models, the Holt (1957) and Winters (1960), because Makridakis and Hibon (2000) report that simple, seasonal exponential smoothing models with seasonality continue to outperform more advanced time series models for large economic time series. Moreover, Makridakis and Hibon (2000) report that equally weighted composite forecasts outperform individual forecasts, a conclusion consistent with Winkler and Makridakis (1983) and Makridakis et al. (1984).

Granger and Newbold (1977) restate the forecast and realization problem. The series to be analyzed and forecast has a fixed mean and variance:

$$E(x_t) = \mu_x$$
$$E(x - \mu_x)^2 = \sigma_x^2.$$

The predictor series, f_2, has mean, f_x, variance σ_x^2, and a correlation ρ with x. The expected squared forecast error is

$$E(x_t - f_t)^2 = (\mu_f - \mu_x)^2 + (\sigma_f - \rho\sigma_x)^2 + (1 - \rho^2)\sigma_x^2. \tag{1.19}$$

A large correlation, ρ, minimizes the expected squared error. If

$$\mu_f = \mu_x \text{ and } \sigma_f = \rho\sigma_x,$$

then for optimal forecasts, the variance of the predictor series is less than the variance of the actual series. The population correlation coefficient is a measure of forecast quality. Granger and Newbold (1977) stated that it is "trivially easy" to obtain a predictor series "highly correlated" with the level of any economic time series.

Granger and Newbold (1977) restated Theil's decomposition of average squared forecast errors. Defining:

$$D_N^2 = \frac{1}{T}\sum_{t=1}^{T}(x_t - f_t)^2 = (\bar{f} - \bar{x})^2 + (s_f - s_x)^2 + 2(1 - r)s_f s_x \tag{1.20}$$

and

$$D_N^2 = (\bar{f} - \bar{x})^2 + (s_f - rs_x)^2 + (1 - r^2)s_x^2. \tag{1.21}$$

If $\bar{f}$ and $\bar{x}$ are sample means of the predictor and predicted series, s_f and s_x are the respective sample standard deviations, and r is the sample correlation coefficient of x and f.

$$U^m = \frac{(\bar{f} - \bar{x})^2}{D_N^2}, \quad U^s = \frac{(s_f - s_x)^2}{D_N^2},$$

$$U^c = 2(1 - r)s_f s_x \big/ D_N^2.$$

As with Theil, $U^M + U^S + U^C = 1$.

If x is a first-order autoregressive process,

$$x_t = ax_{t-1} + \varepsilon_t.$$

An optimal forecast, $f_t = ax_{t-1}$, produces $U^M = 0$, and $U^S + U^C = 1$. A high correlation between predictor and predicted series will most likely not be achieved. The standard deviation of the forecast series is less than the actual series and U^S is substantially different from zero. Granger and Newbold suggest testing for randomness of forecast errors.

Cragg and Malkiel (1968) created a database of five forecasters of long-term earnings forecasts for 185 companies in 1962 and 1963. These five forecast firms included two New York City banks (trust departments), an investment banker, a mutual fund manager, and the final firm was a broker and an investment advisor. The Cragg and Malkiel (1968) forecasts were 5-year average annual growth rates. The earnings forecasts were highly correlated with one another; the highest paired correlation was 0.889 (in 1962) and the lowest paired correlation was 0.450 (in 1963) with most correlations exceeding 0.7. Cragg and Malkiel examined the earnings forecasts among eight "sectors" and found smaller correlation coefficients among the paired correlations within sectors. The correlations of forecasts for 1963 were very highly correlated with 1962 forecasts, exceeding 0.88, for the forecasters. Furthermore, Cragg and Malkiel found that the financial firms' forecasts of earnings were lowly correlated, 0.17–0.45, with forecasts created from time series regressions of earnings over time. Cragg and Malkiel (1968) used the TIC (1966) to measure the efficiency of the financial forecasts and found that the correlations of predicted and realized earnings growth were low, although most were statistically greater than zero. The TICs were large, according to Cragg and Malkiel (1968), although they were less than 1.0 (showing better than no-change forecasting). The TICS were lower (better) within sectors; the forecasts in electronics and electric utility firms were best and foods and oils were the worst firms to forecast earnings growth. Cragg and Malkiel (1968) concluded that their forecasts were little better than past growth rates and that market price-earnings multiples were little better predictors of growth than the financial analysts' forecasts. The Cragg and Malkiel (1968) study was one of the first and most-cited studies of earnings forecasts.

Elton and Gruber (1972) built upon the Cragg and Malkiel study and found similar results. That is, a simple exponentially weighted moving average was a better forecasting model of annual earnings than additive or multiplicative exponential smoothing models with trend or regression models using time as an independent variable. Indeed, a very good model was a naïve model, which assumed a no-change in annual eps with the exception of the prior change that had occurred in earnings. One can clearly see the random walk with drift concept of earnings in the Elton and Gruber (1972). Elton and Gruber compared the naïve and time series forecasts to three financial service firms, and found for their 180 firm sample that two of the three firms were better forecasters than the naïve models. Elton et al. (1981) build upon the Cragg and Malkiel (1968) and Elton and Gruber (1972) results and create an earnings forecasting database that evolves to include over 16,000 companies, the Institutional Brokerage Estimation Services, Inc. (I/B/E/S). Elton et al. (1981) find that earnings revisions, more than the earnings forecasts, determine the securities that will outperform the market. Guerard and Stone (1992) found that the I/B/E/S consensus forecasts were not statistically different than random walk with drift time series forecasts for 648 firms during the 1982–1985 period. Guerard and Stone ran annual eps forecast regressions for rationality and rejected the null hypothesis that analysts' forecasts were rational. Analysts' forecasts were optimistic, producing negative intercepts in the rationality regressions. Analysts' forecasts became less biased during the year and by the third quarter of the year, the bias was essentially zero. Analysts' forecasts were highly correlated with the time series forecasts and latent root regression, used in Chapter 4, reduced forecasting errors in composite earnings forecasting models. Lim (2001), using the I/B/E/S Detailed database from 1984 to December 1996, found forecast bias associated with small and more volatile stocks, experienced poor past stock returns, and had prior negative earnings surprises. Moreover, Lim (2001) found that relative bias was negatively associated with the size of the number of analysts in the brokerage firm. That is, smaller firms with fewer analysts, often with more stale data, produced more optimistic forecasts. Keane and Runkle

(1998) found during the 1983–1991 period that analysts' forecasts were rational, once discretionary special charges are removed. The Keane and Runkle (1998) study is one of the very few studies finding rationality of analysts' forecasts; most find analysts to be optimistic. Further work by Wheeler (1994) will find that firms where analysts agree with the direction of earnings revisions, denoted breadth, will outperform stocks with lesser agreement of earnings revisions. Guerard et al. (1997) combined the work of Elton et al. (1981) and Wheeler (1994) to create a better earnings forecasting model, CTEF, which we use in Chapters 6 and 7. The CTEF variable continues to produce statistically significant excess return in backtest and in identifying real-time security mispricing.

Appendix

Exponential Smoothing

The most simple forecast of a time series can be estimated from an arithmetic mean of the data Davis and Nelson (1937). If one defines *f* as frequencies, or occurrences of the data, and *x* as the values of the series, then the arithmetic mean is

$$A = \frac{f_1x_1 + f_2x_2 + f_3x_3 + \ldots + f_tx_t}{T} \tag{1.22}$$

where $T = f_1 + f_2 + f_3 + \ldots + f_t$.

$$A = \frac{\Sigma f_i x_i}{T}.$$

Alternatively,

$$\frac{\Sigma f_i(x_i - x)}{T}$$

$$A = x + \frac{\Sigma f_i(x_i - x)}{T}. \tag{1.23}$$

The first moment, mean, is

$$A = \frac{\Sigma f_i x_i}{T} = \frac{m_1}{m_0}$$

$$m_0 = \sum f_i = T, m_1 = \sum f_i x_i.$$

If $x = 0$, then

$$\sigma^2 = \frac{\Sigma f_i x_i^2}{T} - A^2$$

$$\sigma^2 = \frac{m_2}{m_o} - \frac{m_1^2}{m_o^2} = (m_0 m_2 - m_1^2)m_0^2. \tag{1.24}$$

Time series models often involve trend, cycle seasonal, and irregular components, Brown (1963). An upward-moving or increasing series over time could be modeled as

$$x_t = a + bt, \tag{1.25}$$

where a is the mean and b is the trend, or rate at which the series increases over time, t. Brown (1963, p. 61) uses the closing price of IBM common stock as his example of an increasing series. One could use a quadrant term, c. If c is positive, then the series

$$x_t = a + bt + ct^2 \tag{1.26}$$

trend is changing toward an increasing trend, whereas a negative c denotes a decreasing rate of trend, from upward to downward.

In an exponential smoothing model, the underlying process is locally constant, $x_t = a$, plus random noise, ε_t.

$$x_t = a\varepsilon_t. \tag{1.27}$$

The average value of $\varepsilon = 0$.

A moving average can be estimated over a portion of the data:

$$M_t = \frac{x_1 + x_{t-1} + \ldots + x_{t-N} + 1}{N}, \tag{1.28}$$

where M_t is the actual average of the most recent N observations.

$$M_t = M_{t-1} + \frac{x_t - x_{t-N}}{N}. \tag{1.29}$$

An exponential smoothing forecast builds upon the moving average concept.

$$s_t(x) = \alpha x_t + (1 - \alpha)s_{t-1}(x),$$

where $\alpha =$ smoothing constant, which is similar to the fraction $1/T$ in a moving average.

$$\begin{aligned} s_t(x) &= \alpha x_t + (1-\alpha)[\alpha x_{t-1} + (1-\alpha)s_{t-2}(x)] \\ &= \alpha \Sigma_{ko}^{t-1}(1-\alpha)^k x_{t-k} + (1-\alpha)^t x_o, \end{aligned} \tag{1.30}$$

where $s_t(x)$ is a linear combination of all past observations. The smoothing constant must be estimated. In a moving average process, the N most recent observations are weighted (equally) by $1/N$ and the average age of the data is

$$k = \frac{0+1+2+\ldots+N-1}{N} = \frac{N-1}{2}.$$

An N-period moving average is equivalent to an exponential smoothing model having an average age of the data. The one-period forecast for an exponential smoothing model is

$$F_{t+1} = F_t + \alpha(y_t - F_t), \tag{1.31}$$

where α is the constant, $0 < \alpha < 1$.

Intuitively, if α is near zero, then the forecast is very close to the previous value's forecast. Alternatively,

$$\begin{aligned} F_{t+1} &= \alpha y_t + (1-\alpha)F_t \\ F_{t+1} &= \alpha y_t + \alpha(1-\alpha)y_{t-1} + (1-\alpha)2F_{t-1}. \end{aligned} \tag{1.32}$$

Makridakis, Wheelwright and Hyndman (1998) express F_{t-1} in terms of F_{t-2} and, over time,

$$\begin{aligned} F_{t-1} = {} & \alpha y_t + \alpha(1-\alpha)y_{t-1} + \alpha(a-\alpha)^2 y_{t-2} + a(1-\alpha)^3 y_{t-3} \\ & + \alpha(1-\alpha)^4 y_{t-4} + \alpha(1-\alpha)^5 y_{t-5} + \ldots \\ & + \alpha(1-\alpha)^{t-1} y_t + (1-\alpha)^t F_1. \end{aligned} \tag{1.33}$$

Different values of α produce different mean squared errors. If one sought to minimize the mean absolute percentage error, the adaptive exponential smoothing can be rewritten as

$$F_{t+1} = \alpha y_t + (1-\alpha)F_t \tag{1.34}$$

$$\alpha t + 1 = \left|\frac{A_t}{M_t}\right|,$$

where

$$\begin{aligned} A_t &= \beta E_t + (1-\beta)A_{t-1} \\ M_t &= \beta|E_t| + (1-\beta)M_{t-1} \\ E_t &= y_t - F_t. \end{aligned}$$

A_t is a smoothed estimate of the forecast error and a weighted average of A_{t-1} and the last forecast error, E_t.

One of the great forecasting models is the Holt (1957) model that allowed forecasting of data with trends. Holt's linear exponential smoothing forecast is

$$\begin{aligned} L_t &= \alpha y_t + (1-\alpha)(L_{t-1} + b_{t-1}) \\ b_t &= \beta(L_t - L_{t-1}) + (1-\beta)b_{t-1} \\ F_{t+m} &= L_t + b_t m. \end{aligned} \quad (1.35)$$

L_t is the level of the series at time t, and b_t is the estimate of the slope of the series at time t. The Holt model forecast should be better forecasts than adaptive exponential smoothing models, which lack trends. Makridakis et al. (1998) remind the reader that the Holt model is often referred to as "double exponential smoothing." If $\alpha = \beta$, then the Holt model is equal to Brown's double exponential smoothing model.

The Hold (1957) and Winters (1960) seasonal model can be written as

$$\text{(Level)} \quad L_t = \alpha \frac{y_t}{s_{t-s}} + (1-\alpha)(L_{t-1} + b_{t-1})$$

$$(Trend) \quad b_t + \beta(L_t - L_{t-1}) + (a - \beta)b_{t-1}$$

$$(Seasonal) \quad s_t = \gamma \frac{y_t}{L_t} + (a - \gamma)s_{t-s}$$

$$(Forecast) \quad F_{t+m} = (L_t + b_t m)S_{t-s+m}.$$

Seasonality is the number of months or quarters, L_t is the level of the series, b_t is the trend of the series, and s_t is the seasonal component.

References

Brown, R.G. 1963. *Smoothing, Forecasting and Prediction of Discrete Time Series*. Englewood Cliffs, N.J.: Prenctice-Hall, Inc.

Clements, M.P. and D.F. Hendry. 1998. *Forecasting Economic Time Series*. Cambridge: Cambridge University Press.

Cragg, J.G. and B.G. Malkiel.1968. "The Consensus and Accuracy of Some Predictions of the Growth of Corporate Earnings". *Journal of Finance* 23, 67–84.

Davis, H.T. and W.F.C. Nelson. 1937. *Elements of Statistics with Application to Economic Data*. Bloomington, Ind: The Principia Press, Inc.

Elton, E.J. and M.J. Gruber. 1972. "Earnings Estimates and the Accuracy of Expectational Data. *Management Science* 18, B409–B424.

Elton, E.J., M.J. Gruber, S.J. Brown, and W.N. Goetzmann. 2009. *Modern Portfolio Theory and Investment Analysis*. New York: Wiley. 7th Edition.

Elton, E. J., M.J. Gruber, and M. Gultekin. 1981. "Expectations and Share Prices." *Management Science* 27, 975–987.

Granger, C.W.J. 1980a. "Testing for Causality: A Personal Viewpoint." *Journal of Economic Dynamics and Control* 2, 329–352.

Granger, C.W.J. 1980b. *Foreccasting in Business and Economics*. New York: Academic Press.

Granger, C.W.J. and P. Newbold. 1977. *Forecasting Economic Time Series*. New York: Academic Press.

Guerard, J.B.,Jr. and B.K. Stone. 1992. "Composite Forecasting of Annual corporate Earnings". In A. Chen, Ed. *Research in Finance* 10, 205–230.

Guerard, J.B., Jr., M. Gultekin, and B.K. Stone. 1997. "The Role of Fundamental Data and Analysts' Earnings Breadth, Forecasts, and Revisions in the Creation of Efficient Portfolios." In *Research in Finance* 15, edited by A. Chen.

Holt, C.C. 1957. "Forecasting Trends and Seasonals by Exponentially Weighted Moving Averages" O.N.R. Memorandum, No. 52, Pittsburgh: Carnegie Institute of Technology. Holt's unpublished paper was published, with several corrected two graphical errors in the *International Journal of Forecasting* 20(2004), 5–10, The paper was published due to its successful use in the M1 competition, see Makridakis et al. (1984), and M3 competition see Makridakis and Hibon (2000).

Holt's unpublished paper was published, with several corrected two graphical errors in the *International Journal of Forecasting* 20(2004), 5–10.

Keane, M.P. and D.E. Runkle. 1998. "Are Financial Analysts' Forecasts of Corporate Profits Rational?" *The Journal of Political Eco*nomy 106, 768–805.

Klein, L. 1985. *Economic Theory and Econometrics*. Philadelphia: University of Pennsylvania Press. Chapter 22.

Lim, T. 2001. "Rationality and Analysts' Forecast Bias". *Journal of Finance* 56, 369–385.

Makridakis, S. and M. Hibon. 2000. "The M3-Competition: Results, Conclusions and Implications." *International Journal of Forecasting* 16, 451–476.

Makridakis, S., S.C. Wheelwright, and R. J. Hyndman. 1998. *Forecasting: Methods and Applications*, Third edition. New York: John Wiley & Sons. Chapters 7 and 8.

Makridakis, S. and Winkler, R.L. 1983. "Averages of Forecasts: Some Empirical Results". *Management Science* 29, 987–996.

Makridakis, S., A. Anderson, R. Carbone, R. Fildes, M. Hibon, J. Newton, E. Parzen, and R. Winkler. 1984. *The Forecasting Accuracy of Major Time Series Methods*. New York: Wiley,

Mincer, J. and V. Zarnowitz. (1969). "The Evaluation of Economic Forecasts," J. Mincer editor, *Economic Forecasts and Expectations*, New York, Columbia University Press.

Theil, H. 1961. *Economic Forecasts and Policy*. Amsterdam: North-Holland Publishing Company. Second Revised Edition.

Theil, H. 1966. *Applied Economic Forecasting*. Amsterdam, North-Holland.

Wheeler, L. 1994. "Changes in Consensus Earnings Estimates and their Impact on Stock Returns", in B. Bruce and C. Epstein, Eds., *The Handbook of Corporate Earnings Analysis*. Chicago: Probus Publishing Company.

Winkler, R.L. and Makridakis, S. 1983. "The Combination of Forecasts." *Journal of the Royal Statistical Society*, Series A., 146, 150–157.

Winters, P.R. 1960. "Forecasting Sales by Exponentially Weighted Averages", *Management Science* 6, 324–342.

Chapter 2
Regression Analysis and Forecasting Models

A forecast is merely a prediction about the future values of data. However, most extrapolative model forecasts assume that the past is a proxy for the future. That is, the economic data for the 2012–2020 period will be driven by the same variables as was the case for the 2000–2011 period, or the 2007–2011 period. There are many traditional models for forecasting: exponential smoothing, regression, time series, and composite model forecasts, often involving expert forecasts. Regression analysis is a statistical technique to analyze quantitative data to estimate model parameters and make forecasts. We introduce the reader to regression analysis in this chapter.

The horizontal line is called the *X*-axis and the vertical line the *Y*-axis. Regression analysis looks for a relationship between the *X* variable (sometimes called the "independent" or "explanatory" variable) and the *Y* variable (the "dependent" variable).

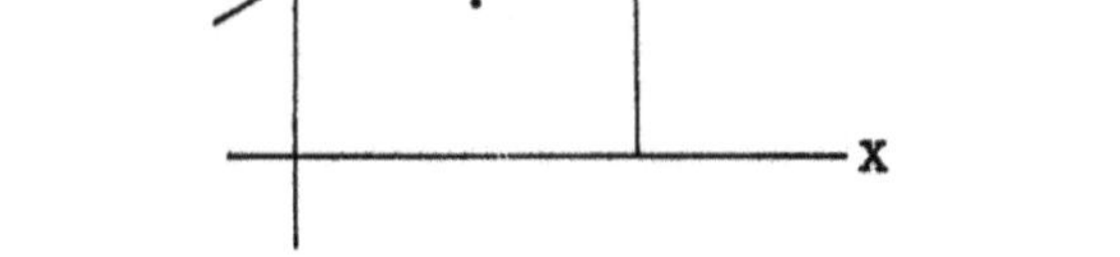

For example, *X* might be the aggregate level of personal disposable income in the United States and *Y* would represent personal consumption expenditures in the United States, an example used in Guerard and Schwartz (2007). By looking up these numbers for a number of years in the past, we can plot points on the graph. More specifically, regression analysis seeks to find the "line of best fit" through the points. Basically, the regression line is drawn to best approximate the relationship

J.B. Guerard, Jr., *Introduction to Financial Forecasting in Investment Analysis*,
DOI 10.1007/978-1-4614-5239-3_2, © Springer Science+Business Media New York 2013

between the two variables. Techniques for estimating the regression line (i.e., its intercept on the Y-axis and its slope) are the subject of this chapter. Forecasts using the regression line assume that the relationship which existed in the past between the two variables will continue to exist in the future. There may be times when this assumption is inappropriate, such as the "Great Recession" of 2008 when the housing market bubble burst. The forecaster must be aware of this potential pitfall. Once the regression line has been estimated, the forecaster must provide an estimate of the future level of the independent variable. The reader clearly sees that the forecast of the independent variable is paramount to an accurate forecast of the dependent variable.

Regression analysis can be expanded to include more than one independent variable. Regressions involving more than one independent variable are referred to as multiple regression. For example, the forecaster might believe that the number of cars sold depends not only on personal disposable income but also on the level of interest rates. Historical data on these three variables must be obtained and a plane of best fit estimated. Given an estimate of the future level of personal disposable income and interest rates, one can make a forecast of car sales.

Regression capabilities are found in a wide variety of software packages and hence are available to anyone with a microcomputer. Microsoft Excel, a popular spreadsheet package, SAS, SCA, RATS, and EViews can do simple or multiple regressions. Many statistics packages can do not only regressions but also other quantitative techniques such as those discussed in Chapter 3 (Time Series Analysis and Forecasting). In simple regression analysis, one seeks to measure the statistical association between two variables, X and Y. Regression analysis is generally used to measure how changes in the independent variable, X, influence changes in the dependent variable, Y. Regression analysis shows a statistical association or correlation among variables, rather than a causal relationship among variables.

The case of simple, linear, least squares regression may be written in the form

$$Y = \alpha + \beta X + \varepsilon, \tag{2.1}$$

where Y, the dependent variable, is a linear function of X, the independent variable. The parameters α and β characterize the population regression line and ε is the randomly distributed error term. The regression estimates of α and β will be derived from the principle of least squares. In applying least squares, the sum of the squared regression errors will be minimized; our regression errors equal the actual dependent variable minus the estimated value from the regression line. If Y represents the actual value and Y the estimated value, then their difference is the error term, e. Least squares regression minimized the sum of the squared error terms. The simple regression line will yield an estimated value of Y, $\hat{Y}$, by the use of the sample regression:

$$\hat{Y} = a + \beta X. \tag{2.2}$$

In the estimation (2.2), a is the least squares estimate of α and b is the estimate of β. Thus, a and b are the regression constants that must be estimated. The least squares regression constants (or statistics) α and β are unbiased and efficient (smallest variance) estimators of α and β. The error term, e_i, is the difference between the actual and estimated dependent variable value for any given independent variable values, X_i.

$$e_i = \hat{Y}_i - Y_i. \tag{2.3}$$

The regression error term, e_i, is the least squares estimate of ε_i, the actual error term.[1]

To minimize the error terms, the least squares technique minimizes the sum of the squares error terms of the N observations,

$$\sum_{i=1}^{N} e_i^2. \tag{2.4}$$

The error terms from the N observations will be minimized. Thus, least squares regression minimizes:

$$\sum_{i=1}^{N} e_i^2 = \sum_{i=1}^{N} [Y_i - \hat{Y}_i]^2 = \sum_{i=1}^{N} [Y_i - (\alpha + bX_i)]^2. \tag{2.5}$$

To assure that a minimum is reached, the partial derivatives of the squared error terms function

$$\sum_{i=1}^{N} = [Y_i - (\alpha + bX_i)]^2$$

will be taken with respect to a and b.

$$\begin{aligned}\frac{\partial \sum_{i=1}^{N} e_i^2}{\partial a} &= 2\sum_{i=1}^{N} (Y_i - a - bX_i)(-1) \\ &= -2\left(\sum_{i=1}^{N} Y_i - \sum_{i=1}^{N} a - b\sum_{i=1}^{N} X_i\right)\end{aligned}$$

[1] The reader is referred to an excellent statistical reference, S. Makridakis, S.C. Wheelwright, and R. J. Hyndman, *Forecasting: Methods and Applications*, Third Edition (New York; Wiley, 1998), Chapter 5.

$$\frac{\partial \sum_{i=1}^{N} e_i^2}{\partial b} = 2\sum_{i=1}^{N}(Y_i - a - bX_i)(-Xi)$$
$$= -2\left(\sum_{i=1}^{N} Y_i X_i - \sum_{i=1}^{N} X_i - b\sum_{i=1}^{N} X_1^2\right).$$

The partial derivatives will then be set equal to zero.

$$\frac{\partial \sum_{i=1}^{N} e_i^2}{\partial a} = -2\left(\sum_{i=1}^{N} Y_i - \sum_{i=1}^{N} a - b\sum_{i=1}^{N} X_i\right) = 0$$
$$\frac{\partial \sum_{i=1}^{N} e_i^2}{\partial b} = -2\left(\sum_{i=1}^{N} YX_i - \sum_{i=1}^{N} X_l - b\sum_{i=1}^{N} X_1^2\right) = 0. \tag{2.6}$$

Rewriting these equations, one obtains the normal equations:

$$\sum_{i=1}^{N} Y_i = \sum_{i=1}^{N} a + b\sum_{i=1}^{N} X_i$$
$$\sum_{i=1}^{N} Y_i X_i = a\sum_{i=1}^{N} X_i + b\sum_{i=1}^{N} X_1^2. \tag{2.7}$$

Solving the normal equations simultaneously for a and b yields the least squares regression estimates:

$$\hat{a} = \frac{\left(\sum_{i=1}^{N} X_i^2\right)\left(\sum_{i=1}^{N} Y_i\right) - \left(\sum_{i=1}^{N} X_i Y_i\right)}{N\left(\sum_{i=1}^{N} X_i^2\right) - \left(\sum_{i=1}^{N} X_i\right)^2},$$
$$\hat{b} = \frac{\left(\sum_{i=1}^{N} X_i Y_i\right) - \left(\sum_{i=1}^{N} X_i\right)\left(\sum_{i=1}^{N} Y_i\right)}{N\left(\sum_{i=1}^{N} X_i^2\right) - \left(\sum_{i=1}^{N} X_i\right)^2}. \tag{2.8}$$

An estimation of the regression line's coefficients and goodness of fit also can be found in terms of expressing the dependent and independent variables in terms of deviations from their means, their sample moments. The sample moments will be denoted by M.

$$M_{XX} = \sum_{i=1}^{N} x_i^2 = \sum_{i=1}^{N} (x_i - \bar{x})^2$$
$$= N \sum_{i=1}^{N} X_i - \left(\sum_{i=1}^{N} X_i \right)^2$$

$$M_{XY} = \sum_{i=1}^{N} x_i y_i = \sum_{i=1}^{N} (X_i - \bar{X})(Y_i - \bar{Y})$$
$$= N \sum_{i=1}^{N} X_i Y_i - \left(\sum_{i=1}^{N} X_i \right) \left(\sum_{i=1}^{N} Y_i \right)$$

$$M_{YY} = \sum_{i=1}^{N} y_i^2 = \sum_{i=1}^{N} (Y - \bar{Y})^2$$
$$= N \left(\sum_{i=1}^{N} Y_i^2 \right) - \sum_{i=1}^{N} (Y_i)^2.$$

The slope of the regression line, b, can be found by

$$b = \frac{M_{XY}}{M_{XX}} \tag{2.9}$$

$$a = \frac{\sum_{i=1}^{N} Y_i}{N} - b \frac{\sum_{i=1}^{N} X_i}{N} = \bar{y} - b\bar{X}. \tag{2.10}$$

The standard error of the regression line can be found in terms of the sample moments.

$$S_e^2 = \frac{M_{XX}(M_{YY}) - (M_{XY})^2}{N(N-2)M_{XX}}$$
$$S_e = \sqrt{S_e^2}. \tag{2.11}$$

The major benefit in calculating the sample moments is that the correlation coefficient, r, and the coefficient of determination, r^2, can easily be found.

$$r = \frac{M_{XY}}{(M_{XX})(M_{YY})}$$
$$R^2 = (r)^2. \tag{2.12}$$

The coefficient of determination, R^2, is the percentage of the variance of the dependent variable explained by the independent variable. The coefficient of determination cannot exceed 1 nor be less than zero. In the case of $R^2 = 0$, the regression line's $Y = Y$ and no variation in the dependent variable are explained. If the dependent variable pattern continues as in the past, the model with time as the independent variable should be of good use in forecasting.

The firm can test whether the a and b coefficients are statistically different from zero, the generally accepted null hypothesis. A t-test is used to test the two null hypotheses:

$H_{0_1}: a = 0$
$H_{A_1}: a \text{ ne } 0$
$H_{0_2}: \beta = 0$
$H_{A_2}: \beta \text{ ne } 0,$
where ne denotes not equal.

The H_0 represents the null hypothesis while H_A represents the alternative hypothesis. To reject the null hypothesis, the calculated t-value must exceed the critical t-value given in the t-tables in the appendix. The calculated t-values for a and b are found by

$$t_a = \frac{a - \alpha}{S_e}\sqrt{\frac{N(M_{XX})}{M_{XX} + (N\bar{X})^2}}$$
$$t_b = \frac{b - \beta}{S_e}\sqrt{\frac{(M_{XX})}{N}}. \tag{2.13}$$

The critical t-value, t_c, for the 0.05 level of significance with $N - 2$ degrees of freedom can be found in a t-table in any statistical econometric text. One has a statistically significant regression model if one can reject the null hypothesis of the estimated slope coefficient.

We can create 95% confidence intervals for a and b, where the limits of a and b are

$$a + ta/_2 S_e{}^{+}\sqrt{\frac{(N\bar{X})^2 + M_{XX}}{N(M_{XX})}}$$
$$b + ta/_2 S_e\sqrt{\frac{N}{M_{XX}}}. \tag{2.14}$$

To test whether the model is a useful model, an F-test is performed where

$H_0 = \alpha = \beta = 0$
$H_A = \alpha \text{ ne } \beta \text{ ne } 0$

$$F = \frac{\sum_{i=1}^{N} Y^2 \div 1 - \beta^2 \sum_{i=1}^{N} X_i^2}{\sum_{i=1}^{N} e^2 \div N - 2}. \tag{2.15}$$

As the calculated F-value exceeds the critical F-value with $(1, \mathrm{N} - 2)$ degrees of freedom of 5.99 at the 0.05 level of significance, the null hypothesis must be rejected. The 95% confidence level limit of prediction can be found in terms of the dependent variable value:

$$(a + bX_0) + ta/_2 S_e \sqrt{\frac{N(X_0 - \bar{X})^2}{1 + N + M_{XX}}}. \tag{2.16}$$

Examples of Financial Economic Data

The most important use of simple linear regression as developed in (2.9) and (2.10) is the estimation of a security beta. A security beta is estimated by running a regression of 60 months of security returns as a function of market returns. The market returns are generally the Standard & Poor's 500 (S&P500) index or a capitalization-weighted index, such as the value-weighted Index from the Center for Research in Security Prices (CRSP) at the University of Chicago. The data for beta estimations can be downloaded from the Wharton Research Data Services (WRDS) database. The beta estimation for IBM from January 2005 to December 2009, using monthly S&P 500 and the value-weighted CRSP Index, produces a beta of approximately 0.80. Thus, if the market is expected to increase 10% in the coming year, then one would expect IBM to return about 8%. The beta estimation of IBM as a function of the S&P 500 Index using the SAS system is shown in Table 2.1. The IBM beta is 0.80. The t-statistic of the beta coefficient, the slope of the regression line, is 5.53, which is highly statistically significant. The critical 5% t-value is with 30 degrees of freedom 1.96, whereas the critical level of the t-statistic at the 10% is 1.645. The IBM beta is statistically different from zero. The IBM beta is not statistically different from one; the normalized z-statistical is significantly less than 1. That is, $0.80 - 1.00$ divided by the regression coefficient standard error of 0.144 produces a Z-statistic of -1.39, which is less than the critical level of -1.645 (at the 10% level) or -1.96 (at the 5% critical level). The IBM beta is 0.78 (the corresponding t-statistic is 5.87) when calculated versus the value-weighted CRSP Index.[2]

[2] See Fama, *Foundations of Finance*, 1976, Chapter 3, p. 101–2, for an IBM beta estimation with an equally weighted CRSP Index.

Table 2.1 WRDS IBM Beta 1/2005–12/2009

Dependent variable: ret					
Number of observations read: 60					
Number of observations used: 60					
Analysis of variance					
Source	DF	Sum of squares	Mean square	F-value	Pr > F
Model	1	0.08135	0.08135	30.60	<0.0001
Error	58	0.15419	0.00266		
Corrected total	59	0.23554			
Root MSE	0.05156	R^2	0.3454		
Dependent mean	0.00808	Adjusted R^2	0.3341		
Coeff var	638.12982				
Parameter estimates					
Variable	DF	Parameter estimate	Standard error	t-Value	Pr > /t/
Intercept	1	0.00817	0.00666	1.23	0.2244
Sprtrn	1	0.80063	0.14474	5.53	<0.0001

Table 2.2 An Estimated Consumption Function, 1947–2011

Dependent variable: RPCE				
Method: least squares				
Sample(adjusted): 1,259				
Included observations: 259 after adjusting endpoints				
Variable	Coefficient	Std. error	t-Statistic	Prob.
C	−120.0314	12.60258	−9.524349	0.0000
RPDI	0.933251	0.002290	407.5311	0.0000
R^2	0.998455	Mean dependent var		4,319.917
Adjusted R^2	0.998449	S.D. dependent var		2,588.624
S.E. of regression	101.9488	Akaike info criterion		12.09451
Sum squared resid	2,671,147	Schwarz criterion		12.12198
Log likelihood	−1,564.239	F-statistic		166,081.6
Durbin–Watson stat	0.197459	Prob(F-statistic)		0.000000

Let us examine another source of real-business economic and financial data. The St. Louis Federal Reserve Bank has an economic database, denoted FRED, containing some 41,000 economic series, available at no cost, via the Internet, at http://research.stlouisfed.org/fred2. Readers are well aware that consumption makes up the majority of real Gross Domestic Product, denoted GDP, the accepted measure of output in our economy. Consumption is the largest expenditure, relative to gross investment, government spending, and net exports in GDP data. If we download and graph real GDP and real consumption expenditures from FRED from 1947 to 2011, shown in Chart 2, one finds that real GDP and real consumption expenditures, in 2005 $, have risen substantially in the postwar period. Moreover, there is a highly statistical significant relationship between real GDP and consumption if one estimates an ordinary least squares (OLS) line of the form of (2.8) with real GDP as the dependent variable and real consumption as the independent variable. The reader is referred to Table 2.2.

Table 2.3 An estimated consumption function, with lagged income

Dependent variable: RPCE
Method: least squares
Sample(adjusted): 2,259
Included observations: 258 after adjusting endpoints

Variable	Coefficient	Std. error	*t*-Statistic	Prob.
C	−118.5360	12.73995	−9.304274	0.0000
RPDI	0.724752	0.126290	5.738800	0.0000
LRPDI	0.209610	0.126816	1.652864	0.0996
R^2	0.998470	Mean dependent var		4,332.278
Adjusted R^2	0.998458	S.D. dependent var		2,585.986
S.E. of regression	101.5529	Akaike info criterion		12.09060
Sum squared resid	2,629,810	Schwarz criterion		12.13191
Log likelihood	−1,556.687	*F*-statistic		83,196.72
Durbin–Watson stat	0.127677	Prob(*F*-statistic)		0.000000

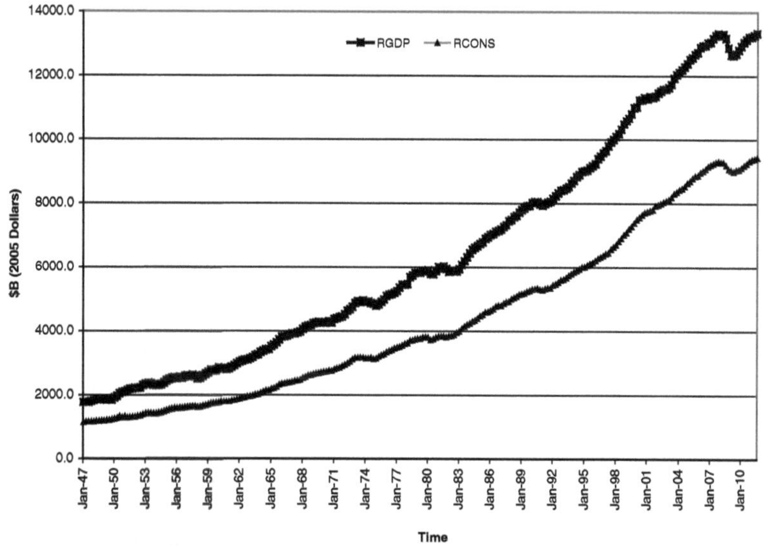

Source: US Department of Commerce, Bureau of Economic Analysis, Series GDPC1 and PCECC96, 1947–2011, seasonally-adjusted, Chained 2005 Dollars

The slope of consumption function is 0.93, and is highly statistically significant.[3]

The introduction of current and lagged income variables in the consumption function regression produces statistically significant coefficients on both current and lagged income, although the lagged income variable is statistically significant at the 10% level. The estimated regression line, shown in Table 2.3, is highly statistically significant.

[3] In recent years the marginal propensity to consume has risen to the 0.90 to 0.97 range, see Joseph Stiglitz, *Economics*, 1993, p.745.

Table 2.4 An estimated consumption function, with twice-lagged consumption

Dependent variable: RPCE
Method: least squares
Included observations: 257 after adjusting endpoints

Variable	Coefficient	Std. error	*t*-Statistic	Prob.
C	−120.9900	12.92168	−9.363331	0.0000
RPDI	0.736301	0.126477	5.821607	0.0000
LRPDI	0.229046	0.177743	1.288633	0.1987
L2RPDI	−0.030903	0.127930	−0.241557	0.8093

R^2	0.998474	Mean dependent var	4,344.661
Adjusted R^2	0.998456	S.D. dependent var	2,583.356
S.E. of regression	101.5049	Akaike info criterion	12.09353
Sum squared resid	2,606,723	Schwarz criterion	12.14877
Log likelihood	−1,550.019	F-statistic	55,188.63
Durbin–Watson stat	0.130988	Prob(F-statistic)	0.000000

The introduction of current and once- and twice-lagged income variables in the consumption function regression produces statistically significant coefficients on both current and lagged income, although the lagged income variable is statistically significant at the 20% level. The twice-lagged income variable is not statistically significant. The estimated regression line, shown in Table 2.4, is highly statistically significant.

Autocorrelation

An estimated regression equation is plagued by the first-order correlation of residuals. That is, the regression error terms are not white noise (random) as is assumed in the general linear model, but are serially correlated where

$$\varepsilon_t = \rho\varepsilon_{t=1} + U_t, \quad t = 1, 2, \ldots, N \tag{2.17}$$

ε_t = regression error term at time t, ρ = first-order correlation coefficient, and U_t = normally and independently distributed random variable.

The serial correlation of error terms, known as autocorrelation, is a violation of a regression assumption and may be corrected by the application of the Cochrane–Orcutt (CORC) procedure.[4] Autocorrelation produces unbiased, the expected value of parameter is the population parameter, but inefficient parameters. The variances of the parameters are biased (too low) among the set of linear unbiased estimators and the sample t- and F-statistics are too large. The CORC

[4] D. Cochrane and G.H. Orcutt, "Application of Least Squares Regression to Relationships Containing Autocorrelated Error Terms," *Journal of the American Statistical Association*, 1949, 44: 32–61.

procedure was developed to produce the best linear unbiased estimators (BLUE) given the autocorrelation of regression residuals. The CORC procedure uses the information implicit in the first-order correlative of residuals to produce unbiased and efficient estimators:

$$Y_t = \alpha + \beta X_t + \varepsilon_t$$
$$\hat{\rho} = \frac{\sum e_t, e_t - 1}{\sum e_t^2 - 1}.$$

The dependent and independent variables are transformed by the estimated rho, $\hat{\rho}$, to obtain more efficient OLS estimates:

$$Y_t - \rho Y_{t-1} = \alpha(l - \rho) + \beta(X_t - \rho X_{t-1}) + ut. \tag{2.19}$$

The CORC procedure is an iterative procedure that can be repeated until the coefficients converge. One immediately recognizes that as ρ approaches unity the regression model approaches a first-difference model.

The Durbin–Watson, *D–W*, statistic was developed to test for the absence of autocorrelation:

H_0: $\rho = 0$.

One generally tests for the presence of autocorrelation ($\rho = 0$) using the Durbin–Watson statistic:

$$D - W = d = \frac{\sum_{t=2}^{N} (e_t = e_{t-1})^2}{\sum_{t=2}^{N} e_t^2}. \tag{2.20}$$

The *e*s represent the OLS regression residuals and a two-tailed tail is employed to examine the randomness of residuals. One rejects the null hypothesis of no statistically significant autocorrelation if

$$d < d_L \text{ or } d > 4 - d_U,$$

where d_L is the "lower" Durbin–Watson level and d_U is the "upper" Durbin–Watson level.

The upper and lower level Durbin–Watson statistic levels are given in Johnston (1972). The Durbin–Watson statistic is used to test only for first-order correlation among residuals.

$$D = 2(1 - \rho). \tag{2.21}$$

If the first-order correlation of model residuals is zero, the Durbin–Watson statistic is 2. A very low value of the Durbin–Watson statistic, $d < d_L$, indicates

Table 2.5 An estimated consumption function, 1947–2011

Dependent variable: D(RPCE)
Method: least squares
Included observations: 258 after adjusting endpoints

Variable	Coefficient	Std. error	t-Statistic	Prob.
C	22.50864	2.290291	9.827849	0.0000
D(RPDI)	0.280269	0.037064	7.561802	0.0000
R^2	0.182581	Mean dependent var		32.18062
Adjusted R^2	0.179388	S.D. dependent var		33.68691
S.E. of regression	30.51618	Akaike info criterion		9.682113
Sum squared resid	238,396.7	Schwarz criterion		9.709655
Log likelihood	−1,246.993	F-statistic		57.18084
Durbin-Watson stat	1.544444	Prob(F-statistic)		0.000000

positive autocorrelation between residuals and produces a regression model that is not statistically plagued by autocorrelation.

The inconclusive range for the estimated Durbin–Watson statistic is

$$d_{\mathrm{L}}<d<d_{\mathrm{U}} \text{ or } 4-d_{\mathrm{U}}<4-d_{\mathrm{U}}.$$

One does not reject the null hypothesis of no autocorrelation of residuals if $d_{\mathrm{U}} < d < 4 - d_{\mathrm{U}}$.

One of the weaknesses of the Durbin–Watson test for serial correlation is that only first-order autocorrelation of residuals is examined; one should plot the correlation of residual with various time lags

$$\operatorname{corr}\left(e_t, e_{t-k}\right)$$

to identify higher-order correlations among residuals.

The reader may immediately remember that the regressions shown in Tables 2.1–2.3 had very low Durbin–Watson statistics and were plagued by autocorrelation. We first-difference the consumption function variables and rerun the regressions, producing Tables 2.5–2.7. The R^2 values are lower, but the regressions are not plagued by autocorrelation. In financial economic modeling, one generally first-differences the data to achieve stationarity, or a series with a constant standard deviation.

The introduction of current and lagged income variables in the consumption function regression produces statistically significant coefficients on both current and lagged income, although the lagged income variable is statistically significant at the 10% level. The estimated regression line, shown in Table 2.6, is highly statistically significant, and is not plagued by autocorrelation.

The introduction of current and lagged income variables in the consumption function regression produces statistically significant coefficients on both current and lagged income, statistically significant at the 1% level. The estimated regression line, shown in Table 2.5, is highly statistically significant, and is not plagued by autocorrelation.

Table 2.6 An estimated consumption function, with lagged income

Dependent variable: D(RPCE)
Method: least squares
Included observations: 257 after adjusting endpoints

Variable	Coefficient	Std. error	t-Statistic	Prob.
C	14.20155	2.399895	5.917570	0.0000
D(RPDI)	0.273239	0.034027	8.030014	0.0000
D(LRPDI)	0.245108	0.034108	7.186307	0.0000
R^2	0.320314	Mean dependent var		32.23268
Adjusted R^2	0.314962	S.D. dependent var		33.74224
S.E. of regression	27.92744	Akaike info criterion		9.508701
Sum squared resid	198,105.2	Schwarz criterion		9.550130
Log likelihood	−1,218.868	F-statistic		59.85104
Durbin-Watson stat	1.527716	Prob(F-statistic)		0.000000

Table 2.7 An estimated consumption function, with twice-lagged consumption

Dependent variable: D(RPCE)
Method: least squares
Included observations: 256 after adjusting endpoints

Variable	Coefficient	Std. error	t-Statistic	Prob.
C	12.78746	2.589765	4.937692	0.0000
D(RPDI)	0.262664	0.034644	7.581744	0.0000
D(LRPDI)	0.242900	0.034162	7.110134	0.0000
D(L2RPDI)	0.054552	0.034781	1.568428	0.1180
R^2	0.325587	Mean dependent var		32.34414
Adjusted R^2	0.317558	S.D. dependent var		33.76090
S.E. of regression	27.88990	Akaike info criterion		9.509908
Sum squared resid	196,017.3	Schwarz criterion		9.565301
Log likelihood	−1,213.268	F-statistic		40.55269
Durbin–Watson stat	1.535845	Prob(F-statistic)		0.000000

The introduction of current and once- and twice-lagged income variables in the consumption function regression produces statistically significant coefficients on both current and lagged income, although the twice-lagged income variable is statistically significant at the 15% level. The estimated regression line, shown in Table 2.7, is highly statistically significant, and is not plagued by autocorrelation.

Many economic time series variables increase as a function of time. In such cases, a nonlinear least squares (NLLS) model may be appropriate; one seeks to estimate an equation in which the dependent variable increases by a constant growth rate rather than a constant amount.[5] The nonlinear regression equation is

[5] The reader is referred to C.T. Clark and L.L. Schkade, Statistical Analysis for Administrative Decisions (Cincinnati: South-Western Publishing Company, 1979) and Makridakis, Wheelwright, and Hyndman, *Op. Cit.*, 1998, pages 221–225, for excellent treatments of this topic.

$$Y = ab^x$$
$$\text{or } \log\ Y = \log a + \log BX. \tag{2.22}$$

The normal equations are derived from minimizing the sum of the squared error terms (as in OLS) and may be written as

$$\begin{aligned}\sum(\log\ Y) &= N(\log\ a) + (\log\ b)\sum X \\ \sum(X\log\ Y) &= (\log\ a)\sum X + (\log\ b)\sum X^2.\end{aligned} \tag{2.23}$$

The solutions to the simplified NLLS estimation equation are

$$\log\ a = \frac{\sum(\log Y)}{N} \tag{2.24}$$

$$\log\ b = \frac{\sum(X\log Y)}{\sum X^2}. \tag{2.25}$$

Multiple Regression

It may well be that several economic variables influence the variable that one is interested in forecasting. For example, the levels of the Gross National Product (GNP), personal disposable income, or price indices can assert influences on the firm. Multiple regression is an extremely easy statistical tool for researchers and management to employ due to the great proliferation of computer software. The general form of the two-independent variable multiple regression is

$$Y_t = \beta_1 + \beta_2 X_{2t} + \beta_3 X_{3t} + \varepsilon_t, \quad t = 1,\ldots,N. \tag{2.26}$$

In matrix notation multiple regression can be written:

$$Y = X\beta + \varepsilon. \tag{2.27}$$

Multiple regression requires unbiasedness, the expected value of the error term is zero, and the X's are fixed and independent of the error term. The error term is an identically and independently distributed normal variable. Least squares estimation of the coefficients yields

$$\begin{aligned}\hat{\beta} &= (\hat{\beta}_1, \hat{\beta}_2, \hat{\beta}_3) \\ Y &= X\hat{\beta} + e.\end{aligned} \tag{2.28}$$

Multiple regression, using the least squared principle, minimizes the sum of the squared error terms:

$$\sum_{i=1}^{N} e_1^2 = e'e$$
$$(Y - X\hat{\beta})'(Y - X\hat{\beta}). \tag{2.29}$$

To minimize the sum of the squared error terms, one takes the partial derivative of the squared errors with respect to $\hat{\beta}$ and the partial derivative set equal to zero.

$$\partial \frac{(e'e)}{\partial \beta} = -2X'Y + 2X'X\hat{\beta} = 0 \tag{2.30}$$

$$\hat{\beta} = (X'X)^{-1}X'Y.$$

Alternatively, one could solve the normal equations for the two-variable to determine the regression coefficients.

$$\begin{aligned}\sum Y &= \beta_1 N + \hat{\beta}_2 \sum X_2 + \hat{\beta}_3 \sum X_3 \\ \sum X_2 Y &= \hat{\beta}_1 \sum X_2 + \hat{\beta}_2 X_2{}^2 + \hat{\beta}_3 \sum X_3^2 \\ \sum X_3 Y &= \hat{\beta}_1 \sum X_3 + \hat{\beta}_2 \sum X_2 X_3 + \hat{\beta}_3 \sum X_3^2.\end{aligned} \tag{2.31}$$

When we solved the normal equation, (2.7), to find the a and b that minimized the sum of our squared error terms in simple liner regression, and when we solved the two-variable normal equation, equation (2.31), to find the multiple regression estimated parameters, we made several assumptions. First, we assumed that the error term is independently and identically distributed, i.e., a random variable with an expected value, or mean of zero, and a finite, and constant, standard deviation. The error term should not be a function of time, as we discussed with the Durbin–Watson statistic, equation (2.21), nor should the error term be a function of the size of the independent variable(s), a condition known as heteroscedasticity. One may plot the residuals as a function of the independent variable(s) to be certain that the residuals are independent of the independent variables. The error term should be a normally distributed variable. That is, the error terms should have an expected value of zero and 67.6% of the observed error terms should fall within the mean value plus and minus one standard deviation of the error terms (the so-called Bell Curve or normal distribution). Ninety-five percent of the observations should fall within the plus or minus two standard deviation levels, the so-called 95% confidence interval. The presence of extreme, or influential, observations may distort estimated regression lines and the corresponding estimated residuals. Another problem in regression analysis is the assumed independence of the independent variables in equation (2.31). Significant correlations may produce estimated regression coefficients that are "unstable" and have the "incorrect" signs, conditions that we will observe in later chapters. Let us spend some time discussing two problems discussed in this section, the problems of influential observations, commonly known as outliers, and the correlation among independent variables, known as multicollinearity.

There are several methods that one can use to identify influential observations or outliers. First, we can plot the residuals and 95% confidence intervals and examine how many observations have residuals falling outside these limits. One should expect no more than 5% of the observations to fall outside of these intervals. One may find that one or two observations may distort a regression estimate even if there are 100 observations in the database. The estimated residuals should be normally distributed, and the ratio of the residuals divided by their standard deviation, known as standardized residuals, should be a normal variable. We showed, in equation (2.31), that in multiple regression

$$\hat{\beta} = (X'X)X'Y.$$

The residuals of the multiple regression line are given by

$$e = Y' - \hat{\beta}X.$$

The standardized residual concept can be modified such that the reader can calculate a variation on that term to identify influential observations. If we delete observation i in a regression, we can measure the change in estimated regression coefficients and residuals. Belsley et al. (1980) showed that the estimated regression coefficients change by an amount, DFBETA, where

$$\text{DFBETA}_i = \frac{(X'X)^{-1}X'e_i}{1 - h_i}, \tag{2.32}$$

where $h_i = X_i(X'X)^{-1}X_i'$.

The h_i or "hat" term is calculated by deleting observation i. The corresponding residual is known as the studentized residual, sr, and defined as

$$\text{sr}_i = \frac{e_i}{\hat{\sigma}\sqrt{1 - h_i}}, \tag{2.33}$$

where $\hat{\sigma}$ is the estimated standard deviation of the residuals. A studentized residual that exceeds 2.0 indicates a potential influential observation (Belsley et al. 1980). Another distance measure has been suggested by Cook (1977), which modifies the studentized residual, to calculate a scaled residual known as the Cook distance measure, CookD. As the researcher or modeler deletes observations, one needs to compare the original matrix of the estimated residual's variance matrix. The COVRATIO calculation performs this calculation, where

$$\text{COVRATIO} = \frac{1}{\left[\frac{n-p-1}{n-p} + \frac{e_i^*}{(n-p)}\right]^p (1 - h_i)}, \tag{2.34}$$

where n = number of observations, p = number of independent variables, and e_i^* = deleted observations.

If the absolute value of the deleted observation >2, then the COVRATIO calculation approaches

$$1 - \frac{3p}{n}. \tag{2.35}$$

A calculated COVRATIO that is larger than $3p/n$ indicates an influential observation. The DFBETA, studentized residual, CookD, and COVRATIO calculations may be performed within SAS. The identification of influential data is an important component of regression analysis. One may create variables for use in multiple regression that make use of the influential data, or outliers, to which they are commonly referred.

The modeler can identify outliers, or influential data, and rerun the OLS regressions on the re-weighted data, a process referred to as robust (ROB) regression. In OLS all data is equally weighted. The weights are 1.0. In ROB regression one weights the data universally with its OLS residual; i.e., the larger the residual, the smaller the weight of the observation in the ROB regression. In ROB regression, several weights may be used. We will see the Huber (1973) and Beaton-Tukey (1974) weighting schemes in our analysis. In the Huber robust regression procedure, one uses the following calculation to weigh the data:

$$w_i = \left(1 - \left(\frac{|e_i|}{\sigma_i}\right)^2\right)^2, \tag{2.36}$$

where e_i = residual i, σ_i = standard deviation of residual, and w_i = weight of observation i.

The intuition is that the larger the estimated residual, the smaller the weight. A second robust re-weighting scheme is calculated from the Beaton-Tukey biweight criteria where

$$w_i = \begin{cases} \left(1 - \left(\dfrac{\frac{|e_i|}{\sigma_e}}{4.685}\right)^2\right)^2, & \text{if } \dfrac{|e_i|}{\sigma_e} > 4.685; \\ 1, & \text{if } \dfrac{|e_i|}{\sigma_e} < 4.685. \end{cases} \tag{2.37}$$

A second major problem is one of multicollinearity, the condition of correlations among the independent variables. If the independent variables are perfectly correlated in multiple regression, then the $(X'X)$ matrix of (2.31) cannot be inverted and the multiple regression coefficients have multiple solutions. In reality, highly correlated independent variables can produce unstable regression coefficients due to an unstable $(X'X)^{-1}$ matrix. Belsley et al. advocate the calculation of a condition number, which is the ratio of the largest latent root of the correlation matrix relative to the smallest latent root of the correlation matrix. A condition number exceeding 30.0 indicates severe multicollinearity.

The latent roots of the correlation matrix of independent variables can be used to estimate regression parameters in the presence of multicollinearity. The latent roots, $l_1, l_2, \ldots, l_p$ and the latent vectors $\gamma_1, \gamma_2, \ldots, \gamma_p$ of the P independent variables can describe the inverse of the independent variable matrix of (2.29).

$$(X'X)^{-1} = \sum_{j=1}^{p} l_j^{-1} \gamma_j \gamma_j'.$$

Multicollinearity is present when one observes one or more small latent vectors. If one eliminates latent vectors with small latent roots ($l < 0.30$) and latent vectors ($\gamma < 0.10$), the "principal component" or latent root regression estimator may be written as

$$\hat{\beta}_{\text{LRR}} = \sum_{j=0}^{P} f_j \delta_j,$$

where $f_j = \frac{-\eta\gamma_0 {\lambda_j}^{-1}}{\sum_q \gamma_0^2 {\lambda_q}^{-1}}$,

where $n^2 = \Sigma(y - \bar{y})^2$

and λ are the "nonzero" latent vectors. One eliminates the latent vectors with non-predictive multicollinearity. We use latent root regression on the Beaton-Tukey weighted data in Chapter 4.

The Conference Board Composite Index of Leading Economic Indicators and Real US GDP Growth: A Regression Example

The composite indexes of leading (leading economic indicators, LEI), coincident, and lagging indicators produced by The Conference Board are summary statistics for the US economy. Wesley Clair Mitchell of Columbia University constructed the indicators in 1913 to serve as a barometer of economic activity. The leading indicator series was developed to turn upward before aggregate economic activity increased, and decrease before aggregate economic activity diminished.

Historically, the cyclical turning points in the leading index have occurred before those in aggregate economic activity, cyclical turning points in the coincident index have occurred at about the same time as those in aggregate economic activity, and cyclical turning points in the lagging index generally have occurred after those in aggregate economic activity.

The Conference Board's components of the composite leading index for the year 2002 reflects the work and variables shown in Zarnowitz (1992) list, which continued work of the Mitchell (1913 and 1951), Burns and Mitchell (1946), and Moore (1961). The Conference Board index of leading indicators is composed of

1. Average weekly hours (mfg.)
2. Average weekly initial claims for unemployment insurance
3. Manufacturers' new orders for consumer goods and materials
4. Vendor performance
5. Manufacturers' new orders of nondefense capital goods
6. Building permits of new private housing units
7. Index of stock prices
8. Money supply
9. Interest rate spread
10. Index of consumer expectations

The Conference Board composite index of LEI is an equally weighted index in which its components are standardized to produce constant variances. Details of the LEI can be found on The Conference Board Web site, www.conference-board.org, and the reader is referred to Zarnowitz (1992) for his seminal development of underlying economic assumption and theory of the LEI and business cycles (see Table 2.8).

Let us illustrate a regression of real US GDP as a function of current and lagged LEI. The regression coefficient on the LEI variable, 0.232, in Table 2.9, is highly statistically significant because the calculated t-value of 6.84 exceeds 1.96, the 5% critical level. One can reject the null hypothesis of no association between the growth rate of US GDP and the growth rate of the LEI. The reader notes, however, that we estimated the regression line with current, or contemporaneous, values of the LEI series.

The LEI series was developed to "forecast" future economic activity such that current growth of the LEI series should be associated with future US GDP growth rates. Alternatively, one can examine the regression association of the current values of real US GDP growth and previous or lagged values, of the LEI series. How many lags might be appropriate? Let us estimate regression lines using up to four lags of the US LEI series. If one estimates multiple regression lines using the EViews software, as shown in Table 2.10, the first lag of the LEI series is statistically significant, having an estimated t-value of 5.73, and the second lag is also statistically significant, having an estimated t-value of 4.48. In the regression analysis using three lags of the LEI series, the first and second lagged variables are highly statistically significant, and the third lag is not statistically significant because third LEI lag variable has an estimated t-value of only 0.12. The critical

Table 2.8 The conference board leading, coincident, and lagging indicator components

Leading index			Standardization factor
1	BCI-01	Average weekly hours, manufacturing	0.1946
2	BCI-05	Average weekly initial claims for unemployment insurance	0.0268
3	BCI-06	Manufacturers' new orders, consumer goods and materials	0.0504
4	BCI-32	Vendor performance, slower deliveries diffusion index	0.0296
5	BCI-27	Manufacturers' new orders, nondefense capital goods	0.0139
6	BCI-29	Building permits, new private housing units	0.0205
7	BCI019	Stock prices, 500 common stocks	0.0309
8	BCI-106	Money supply, M2	0.2775
9	BCI-129	Interest rate spread, 10-year Treasury bonds less federal funds	0.3364
10	BCI-83	Index of consumer expectations	0.0193
Coincident index			
1	BCI-41	Employees on nonagricultural payrolls	0.5186
2	BCI-51	Personal income less transfer payments	0.2173
3	BCI-47	Industrial production	0.1470
4	BCI-57	Manufacturing and trade sales	0.1170
Lagging index			
1	BCI-91	Average duration of unemployment	0.0368
2	BCI-77	Inventories-to-sales ratio, manufacturing and trade	0.1206
3	BCI-62	Labor cost per unit of output, manufacturing	0.0693
4	BCI-109	Average prime rate	0.2692
5	BCI-101	Commercial and industrial loans	0.1204
6	BCI-95	Consumer installment credit-to-personal income ratio	0.1951
7	BCI-120	Consumer price index for services	0.1886

Table 2.9 Real US GDP and the leading indicators: A contemporaneous examination

Dependent variable: DLOG(RGDP)
Sample(adjusted): 2,210
Included observations: 209 after adjusting endpoints

Variable	Coefficient	Std. error	*t*-Statistic	Prob.
C	0.006170	0.000593	10.40361	0.0000
DLOG(LEI)	0.232606	0.033974	6.846529	0.0000
R^2	0.184638	Mean dependent var		0.007605
Adjusted R^2	0.180699	S.D. dependent var		0.008860
S.E. of regression	0.008020	Akaike info criterion		−6.804257
Sum squared resid	0.013314	Schwarz criterion		−6.772273
Log likelihood	713.0449	*F*-statistic		46.874971
Durbin–Watson stat	1.594358	Prob(*F*-statistic)		0.000000

t-level at the 10% level is 1.645, for 30 observations, and statistical studies often use the 10% level as a minimum acceptable critical level. The third lag is not statistically significant in the three quarter multiple regression analysis. In the four quarter lags analysis of the LEI series, we report that the lag one variable has a *t*-statistic of

Table 2.10 Real GDP and the conference board leading economic indicators

1959 Q1–2011 Q2								
			Lags (LEI)					
Model	Constant	LEI	One	Two	Three	Four	R^2	F-statistic
RGDP	0.006	0.232					0.181	46.875
(t)	10.400	6.850						
RGDP	0.056	0.104	0.218				0.285	42.267
	9.910	2.750	5.730					
RGDP	0.005	0.095	0.136	0.162			0.353	38.45
	9.520	2.600	3.260	4.480				
RGDP	0.005	0.093	0.135	0.164	0.005		0.351	28.679
	9.340	2.530	3.220	3.900	0.120			
RGDP	0.005	0.098	0.140	0.167	−0.041	0.061	0.369	24.862
	8.850	2.680	3.360	4.050	−0.990	1.670		

Table 2.11 The REG procedure

Dependent variable: DLUSGDP
Sample(adjusted): 6,210
Included observations: 205 after adjusting endpoints

Variable	Coefficient	Std. error	t-Statistic	Prob.
C	0.004915	0.000555	8.849450	0.0000
DLOG(LEI)	0.098557	0.036779	2.679711	0.0080
DLOG(L1LEI)	0.139846	0.041538	3.366687	0.0009
DLOG(L2LEI)	0.167168	0.041235	4.054052	0.0001
DLOG(L3LEI)	−0.041170	0.041305	−0.996733	0.3201
DLOG(L4LEI)	0.060672	0.036401	1.666786	0.0971
R^2	0.384488	Mean dependent var		0.007512
Adjusted R^2	0.369023	S.D. dependent var		0.008778
S.E. of regression	0.006973	Akaike info criterion		−7.064787
Sum squared resid	0.009675	Schwarz criterion		−6.967528
Log likelihood	730.1406	F-statistic		24.86158
Durbin–Watson stat	1.784540	Prob(F-statistic)		0.000000

3.36, highly significant; the second lag has a t-statistic of 4.05, which is statistically significant; the third LEI lag variable has a t-statistic of −0.99, not statistically significant at the 10% level; and the fourth LEI lag variable has an estimated t-statistic of 1.67, which is statistically significant at the 10% level. The estimation of multiple regression lines would lead the reader to expect a one, two, and four variable lag structure to illustrate the relationship between real US GDP growth and The Conference Board LEI series. The next chapter develops the relationship using time series and forecasting techniques. This chapter used regression analysis to illustrate the association between real US GDP growth and the LEI series.

The reader is referred to Table 2.11 for EViews output for the multiple regression of the US real GDP and four quarterly lags in LEI.

Table 2.12 The REG procedure model: MODEL1

Dependent variable: dlRGDP
Number of observations read: 209
Number of observations used: 205
Number of observations with missing values: 4

Analysis of variance

Source	DF	Sum of squares	Mean square	F-value	Pr > F
Model	5	0.00604	0.00121	24.85	<0.0001
Error	199	0.00968	0.00004864		
Corrected total	204	0.01572			
	Root MSE	0.00697	R^2	0.3844	
	Dependent mean	0.00751	Adjusted R^2	0.3689	
	Coeff. var	92.82825			

Parameter estimates

Variable	DF	Parameter estimate	Standard error	t-Value	Pr > $\|t\|$	Variance inflation
Intercept	1	0.00492	0.00055545	8.85	<0.0001	0
dlLEI	1	0.09871	0.03678	2.68	0.0079	1.52694
dlLEI_1	1	0.13946	0.04155	3.36	0.0009	1.94696
dlLEI_2	1	0.16756	0.04125	4.06	<0.0001	1.92945
dlLEI_3	1	−0.04121	0.04132	−1.00	0.3198	1.93166
dlLEI_4	1	0.06037	0.03641	1.66	0.0989	1.50421

Collinearity diagnostics

Number	Eigenvalue	Condition index
1	3.08688	1.00000
2	1.09066	1.68235
3	0.74197	2.03970
4	0.44752	2.62635
5	0.37267	2.87805
6	0.26030	3.44367

Proportion of variation

Number	Intercept	dlLEI	dlLEI_1	dlLEI_2	dlLEI_3	dlLEI_4
1	0.02994	0.02527	0.02909	0.03220	0.02903	0.02481
2	0.00016369	0.18258	0.05762	0.00000149	0.06282	0.19532
3	0.83022	0.00047128	0.02564	0.06795	0.02642	0.00225
4	0.12881	0.32579	0.00165	0.38460	0.00156	0.38094
5	0.00005545	0.25381	0.41734	0.00321	0.44388	0.19691
6	0.01081	0.21208	0.46866	0.51203	0.43629	0.19977

We run the real GDP regression with four lags of LEI data in SAS. We report the SAS output in Table 2.12. The Belsley et al. (1980) condition index of 3.4 reveals little evidence of multicollinearity and the collinearity diagnostics reveal no two variables in a row exceeding 0.50. Thus, SAS allows the researcher to specifically address the issue of multicollinearity. We will return to this issue in Chap. 4.

Table 2.13 Modeling dlRGDP by OLS

	Coefficient	Std. error	*t*-Value	*t*-Prob	Part. R^2
Constant	0.00491456	0.0005554	8.85	0.0000	0.2824
dlLEI	0.0985574	0.03678	2.68	0.0080	0.0348
dlLEI_1	0.139846	0.04154	3.37	0.0009	0.0539
dlLEI_2	0.167168	0.04123	4.05	0.0001	0.0763
dlLEI_3	−0.0411702	0.04131	−0.997	0.3201	0.0050
dlLEI_4	0.0606721	0.03640	1.67	0.0971	0.0138
Sigma	0.00697274	RSS	0.00967519164		
R^2	0.384488; $F(5,199) = 24.86$ [0.000]				
Adjusted R^2	0.369023	Log-likelihood	730.141		
No. of observations	205	No. of parameters	6		
Mean(dlRGDP)	0.00751206	S.E.(dlRGDP)	0.00877802		
AR 1–2 test:	$F(2,197) = 3.6873$ [0.0268]*				
ARCH 1–1 test:	$F(1,203) = 1.6556$ [0.1997]				
Normality test:	Chi-squared(2) = 17.824 [0.0001]				
Hetero test:	$F(10,194) = 0.86780$ [0.5644]				
Hetero-X test:	$F(20,184) = 0.84768$ [0.6531]				
RESET23 test:	$F(2,197) = 2.9659$ [0.0538]				

The SAS estimates of the regression model reported in Table 2.12 would lead the reader to believe that the change in real GDP is associated with current, lagged, and twice-lagged LEI.

Alternatively, one could use Oxmetrics, an econometric suite of products for data analysis and forecasting, to reproduce the regression analysis shown in Table 2.13.[6]

An advantage to Oxmetrics is its Automatic Model selection procedure that addresses the issue of outliers. One can use the Oxmetrics Automatic Model selection procedure and find two statistically significant lags on LEI and three outliers: the economically volatile periods of 1971, 1978, and (the great recession of) 2008 (see Table 2.14).

The reader clearly sees the advantage of the Oxmetrics Automatic Model selection procedure.

---

[6] Ox Professional version 6.00 (Windows/U) (C) J.A. Doornik, 1994–2009, PcGive 13.0.See Doornik and Hendry (2009a, b).

Table 2.14 Modeling dlRGDP by OLS

	Coefficient	Std. error	t-Value	t-Prob	Part. R^2
Constant	0.00519258	0.0004846	10.7	0.0000	0.3659
dlLEI_1	0.192161	0.03312	5.80	0.0000	0.1447
dlLEI_2	0.164185	0.03281	5.00	0.0000	0.1118
I:1971-01-01	0.0208987	0.006358	3.29	0.0012	0.0515
I:1978-04-01	0.0331323	0.006352	5.22	0.0000	0.1203
I:2008-10-01	−0.0243503	0.006391	−3.81	0.0002	0.0680
Sigma	0.00633157	RSS	0.00797767502		
R^2	0.49248	$F(5,199) = 38.62$ [0.000]			
Adjusted R^2	0.479728	Log-likelihood	749.915		
No. of observations	205	No. of parameters	6		
Mean(dlRGDP)	0.00751206	se(dlRGDP)	0.00877802		
AR 1–2 test:	$F(2,197) = 3.2141$ [0.0423]				
ARCH 1–1 test:	$F(1,203) = 2.3367$ [0.1279]				
Normality test:	Chi-squared (2) = 0.053943 [0.9734]				
Hetero test:	$F(4,197) = 3.2294$ [0.0136]				
Hetero-X test:	$F(5,196) = 2.5732$ [0.0279]				
RESET23 test:	$F(2,197) = 1.2705$ [0.2830]				

Summary

In this chapter, we introduced the reader to regression analysis and various estimation procedures. We have illustrated regression estimations by modeling consumption functions and the relationship between real GDP and The Conference Board LEI. We estimated regressions using EViews, SAS, and Oxmetrics. There are many advantages with the various regression software with regard to ease of use, outlier estimations, collinearity diagnostics, and automatic modeling procedures. We will use the regression techniques in Chap. 4.

Appendix

Let us follow The Conference Board definitions of the US LEI series and its components:

Leading Index Components

BCI-01 Average weekly hours, manufacturing. The average hours worked per week by production workers in manufacturing industries tend to lead the business cycle because employers usually adjust work hours before increasing or decreasing their workforce.

BCI-05 Average weekly initial claims for unemployment insurance. The number of new claims filed for unemployment insurance is typically more sensitive than either total employment or unemployment to overall business conditions, and this series tends to lead the business cycle. It is inverted when included in the leading index; the signs of the month-to-month changes are reversed, because initial claims increase when employment conditions worsen (i.e., layoffs rise and new hirings fall).

BCI-06 Manufacturers' new orders, consumer goods and materials (in 1996 $). These goods are primarily used by consumers. The inflation-adjusted value of new orders leads actual production because new orders directly affect the level of both unfilled orders and inventories that firms monitor when making production decisions. The Conference Board deflates the current dollar orders data using price indexes constructed from various sources at the industry level and a chain-weighted aggregate price index formula.

BCI-32 Vendor performance, slower deliveries diffusion index. This index measures the relative speed at which industrial companies receive deliveries from their suppliers. Slowdowns in deliveries increase this series and are most often associated with increases in demand for manufacturing supplies (as opposed to a negative shock to supplies) and, therefore, tend to lead the business cycle. Vendor performance is based on a monthly survey conducted by the National Association of Purchasing Management (NAPM) that asks purchasing managers whether their suppliers' deliveries have been faster, slower, or the same as the previous month. The slower-deliveries diffusion index counts the proportion of respondents reporting slower deliveries, plus one-half of the proportion reporting no change in delivery speed.

BCI-27 Manufacturers' new orders, nondefense capital goods (in 1996 $). New orders received by manufacturers in nondefense capital goods industries (in inflation-adjusted dollars) are the producers' counterpart to BCI-06.

BCI-29 Building permits, new private housing units. The number of residential building permits issued is an indicator of construction activity, which typically leads most other types of economic production.

BCI-19 Stock prices, 500 common stocks. The Standard & Poor's 500 stock index reflects the price movements of a broad selection of common stocks traded on the New York Stock Exchange. Increases (decreases) of the stock index can reflect both the general sentiments of investors and the movements of interest rates, which is usually another good indicator for future economic activity.

BCI-106 Money supply (in 1996 $). In inflation-adjusted dollars, this is the M2 version of the money supply. When the money supply does not keep pace with inflation, bank lending may fall in real terms, making it more difficult for the economy to expand. M2 includes currency, demand deposits, other checkable deposits, travelers checks, savings deposits, small denomination time deposits, and balances in money market mutual funds. The inflation adjustment is based on the implicit deflator for personal consumption expenditures.

BCI-129 Interest rate spread, 10-year Treasury bonds less federal funds. The spread or difference between long and short rates is often called the yield curve. This series is constructed using the 10-year Treasury bond rate and the federal funds rate, an overnight interbank borrowing rate. It is felt to be an indicator of the stance of monetary policy and general financial conditions because it rises (falls) when short rates are relatively low (high). When it becomes negative (i.e., short rates are higher than long rates and the yield curve inverts) its record as an indicator of recessions is particularly strong.

BCI-83 Index of consumer expectations. This index reflects changes in consumer attitudes concerning future economic conditions and, therefore, is the only indicator in the leading index that is completely expectations-based. Data are collected in a monthly survey conducted by the University of Michigan's Survey Research Center. Responses to the questions concerning various economic conditions are classified as positive, negative, or unchanged. The expectations series is derived from the responses to three questions relating to (1) economic prospects for the respondent's family over the next 12 months; (2) economic prospects for the Nation over the next 12 months; and (3) economic prospects for the Nation over the next 5 years.

References

Beaton, A.E. and J.W. Tukey, 1974. "The Fitting of Power Series, Meaning Polynomials, Illustrated on Bank-Spectroscopic Data," *Technometrics* 16, 147-185.

Belsley, D.A., E. Kuh, and R.E. Welsch, 1980. *Regression Diagnostics: Identifying Influential Data and Sources of Collinearity*. New York: John Wiley & Sons, Chapter 2.

Burns, A. F. and W.C. Mitchell. 1946. *Measuring Business Cycles*. New York, NBER.

Clements, M. P. and D. F. Hendry. 1998. *Forecasting Economic Time Series*. Cambridge: Cambridge University Press.

Cochrane, D. and G.H. Orcutt. 1949. Application of Least Squares Regression to Relationships Containing Autocorrelated Error Terms. *Journal of the American Statistical Association*. 44:32-61.

Cook, R.D. 1977. "Detection of Influential Observations in Linear Regression." *Technometrics* 19, 15–18.

Doornik, J. A. and D. F. Hendry. 2009a. *Empirical Econometric Modelling*. Timberlake Consultants, Ltd.
Doornik, J. A. and D. F. Hendry. 2009b. *Modelling Dynamic Systems*. Timberlake Consultants, Ltd.
Fama, E.F. 1976. *Foundations of Finance*. New York: Basic Books, Inc., Chapter 3.
Guerard, J.B., Jr. and E. Schwartz, 2007. *Quantitative Corporate Finance*. New York: Springer.
Gunst, R.F. and R.L. Mason, 1980. *Regression Analysis and its Application*, New York: Marcel Dekker, Inc.
Huber, P.J., 1973. "Robust Regression: Asymptotics, Conjectures, and Monte Carlo," *Annals of Statistics* 1, 799-82.
Johnston, J. 1972. *Econometric Methods*. 2nd ed. New York: McGraw-Hill.
Makridakis, S., S.C. Wheelwright, and R. J. Hyndman, 1998. *Forecasting: Methods and Applications*. John Wiley & Sons, 3rd edition. Chapters 5, 6.
Mansfield, E. 1994. *Statistics for Business and Economics*. 5th ed. New York: W.W. Norton &. Company.
Miller, Irwin, and John E. Freund. 1965. *Probability and Statistics for Engineers*. Englewood Cliffs, N.J.: Prentice-Hall.
Mitchell, W.C. 1913. *Business Cycles*, New York, Burt Franklin reprint.
Mitchell, W.C. 1951. *What Happens During Business Cycles: A Progress Report*, New York, NBER.
Moore, G.H. 1961. *Business Cycle Indicators*, 2 volumes, Princeton, Princeton University Press.
Murphy, James L. 1973. *Introductory Econometrics*. Homewood, IL: Richard D. Irwin, Inc.
Nelson, C.R. and C.I. Plosser. 1982. "Trends and Random Walks in Macroeconomic Time Series," *Journal of Monetary Economics*, vol. 10, pp. 139-162.
Stiglitz, Joseph. 1993. *Economics*. New York: W. W. Norton & Company.
Sharpe, W. F. 2007. *Investors and Markets*. Princeton; Princeton University Press, Chapter 4.
Zarnowitz, V. 2004. "The Autonomy of Recent US Growth and Business Cycles". In P. Dua, Ed., *Business Cycles and Economic Growth: An Analysis Using Leading Indicators*. New York: Oxford University Press, 44-82.
Zarnowitz, V. 2001. "The Old and the New in the U.S. Economic Expansion." The Conference Board. EPWP #01-01.
Zarnowitz, V. 1992. *Business Cycles: Theory, History, Indicators, and Forecasting*. Chicago, University of Chicago Press.

Chapter 3
An Introduction to Time Series Modeling and Forecasting

An important aspect of financial decision making may depend on the forecasting effectiveness of the composite index of leading economic indicators, LEI. The leading indicators can be used as an input to a transfer function model of real Gross Domestic Product, GDP. The previous chapter employed four quarterly lags of the LEI series to estimate regression models of association between current rates of growth of real US GDP and the composite index of LEI. This chapter asks the question as to whether changes in forecasted economic indexes help forecast changes in real economic growth. The transfer function model forecasts are compared to several naïve models in terms of testing which model produces the most accurate forecast of real GDP. No-change (NoCH) forecasts of real GDP and random walk with drift (RWD) models may be useful forecasting benchmarks (Mincer and Zarnowitz 1969; Granger and Newbold 1977). Economists have constructed LEI series to serve as a business barometer of the changing US economy since the time of Mitchell (1913). The purpose of this study is to examine the time series forecasts of composite economic indexes produced by The Conference Board (TCB), and test the hypothesis that the leading indicators are useful as an input to a time series model to forecast real output in the United States.

Economic indicators are descriptive and anticipatory time series data used to analyze and forecast changing business conditions. Cyclical indicators are comprehensive series that are systemically related to the business cycle. Business cycles are recurrent sequences of expansions and contractions in aggregate economic activity. Coincident indicators have cyclical movements that approximately correspond with the overall business cycle expansions and contractions. Leading indicators reach their turning points before the corresponding business cycle turns. The lagging indicators reach their turning points after the corresponding turns in the business cycle.

An example of business cycles can be found in the analysis of Irving Fisher (1911), who discussed how changes in the money supply lead to rising prices and an initial fall in the rate of interest, and how this results in raising profits, creating a boom. The interest rate later rises, reducing profits, and ending the boom. A financial crisis ensues when businessmen, whose loan collateral is falling as

J.B. Guerard, Jr., *Introduction to Financial Forecasting in Investment Analysis*,
DOI 10.1007/978-1-4614-5239-3_3,

interest rates rise, run to cash and banks fail. The money supply is one series in TCB index of leading economic indicators, LEI.

Section "ARMA Model Identification in Practice" of this chapter presents an introduction to the models that are estimated and tested in the analysis of the forecasting effectiveness of the leading indicators. Section "Modeling Real GDP: An Example" presents the empirical evidence to support the time series models and reports how models adequately describe the data. Out-of-sample forecasting results are shown in Section "Leading Economic Indicators (LEI) and Real GDP Analysis: The Statistical Evidence, 1970–2002" for the United States and the G7 nations.[1] We present additional evidence on out-of-sample forecasting for the Yen exchange, consumption–income relationship, and Real GDP and LEI transfer function modeling.

Basic Statistical Properties of Economic Series

This chapter develops and forecasts models of economic time series in which we initially use only the past history of the series. The chapter later explores explanatory variables in the forecast models. The time series modeling approach of Box and Jenkins involves the identification, estimation, and forecasting of stationary (or series transformed to stationarity) series through the analysis of the series autocorrelation and partial autocorrelation (PAC) functions.[2] The autocorrelation function examines the correlations of the current value of the economic times series and its previous k-lags. That is, one can measure the correlation of a daily series, of shares, or other assets, by calculating

$$p_{jt} = a + bp_{jt-1}, \tag{3.1}$$

where p_{jt} = today's price of stock j; p_{jt-1} = yesterday's price of stock j; and b is the correlation coefficient.

In a daily shares price series, b is quite large, often approaching a value of 1.00. As the number of lags or previous number of periods increases, the correlation tends to fall. The decrease is usually very gradual.

The PAC function examines the correlation between p_{jt} and p_{jt-2}, holding constant the association between p_{jt} and p_{jt-1}. If a series follows a random walk, the correlation between p_{jt} and p_{jt-1} is one, and the correlation between p_{jt} and p_{jt-2}, holding constant the correlation of p_{jt} and p_{jt-1}, is zero. Random walk series are characterized with decaying autocorrelation functions and a PAC function with a "spike" at lag one, and zeros thereafter. Stationarity implies that the joint probability

[1] Section "ARMA Model Identification in Practice" can be omitted with little loss of continuity with readers more interested in the application of time series models.

[2] This section draws heavily from Box and Jenkins (1970, Chaps. 2 and 3).

$[p(Z)]$ distribution $P(Z_{t1}, Z_{t2})$ is the same for all times t, t_1, and t_2 where the observations are separated by a constant time interval. The autocovariance of a time series at some lag or interval, k, is defined to be the covariance between Z_t and Z_{t+k}:

$$\gamma_k = \operatorname{cov}[Z_t, Z_{t+k}] = E[(Z_t - \mu)(Z_{t+k} - \mu)]. \tag{3.2}$$

One must standardize the autocovariance, as one standardizes the covariance in traditional regression analysis, before one can quantify the statistically significant association between Z_t and Z_{t+k}. The autocorrelation of a time series is the standardization of the autocovariance of a time series relative to the variance of the time series, and the autocorrelation at lag k, ρ_k, is bounded between +1 and −1:

$$\begin{aligned}\rho_k &= \frac{E[(Z_t - \mu)(Z_{t+k} - \mu)]}{\sqrt{E[(Z_t - \mu)^2]E[(Z_{t+k} - \mu)^2]}} \\ &= \frac{E[(Z_t - \mu)(Z_{t+k} - \mu)]}{\sigma_Z^2} = \frac{r_k}{r_0}.\end{aligned} \tag{3.3}$$

The autocorrelation function of the process, $\{\rho_k\}$, represents the plotting of r_k versus time, the lag of k. The autocorrelation function is symmetric about series and thus $\rho_k = \rho_{-k}$; thus, time series analysis normally examines only the positive segment of the autocorrelation function. One may also refer to the autocorrelation function as the correlogram. The statistical estimates of the autocorrelation function are calculated from a finite series of N observations, Z_1, Z_2, Z_3, …, Z_n. The statistical estimate of the autocorrelation function at lag k, r_k, is found by

$$r_k = \frac{C_k}{C_0},$$

where

$$C_k = \frac{1}{N}\sum_{t=1}^{N-k}(Z_t - \bar{Z})(Z_{t+k} - \bar{Z}), \quad k = 0, 1, 2, \ldots, K.$$

C_k is, of course, the statistical estimate of the autocovariance function at lag k. In identifying and estimating parameters in a time series model, one seeks to identify orders (lags) of the time series that are statistically different from zero. The implication of testing whether an autocorrelation estimate is statistically different from zero leads one back to the t-tests used in regression analysis to examine the statistically significant association between variables. One must develop a standard error of the autocorrelation estimate such that a formal t-test can be performed to measure the statistical significance of the autocorrelation estimate. Such a standard error, S_e, estimate was found by Bartlett and, in large samples, is approximated by

$$Var[r_k] \cong \frac{1}{N} \quad \text{and} \quad S_e[r_k] \cong \frac{1}{\sqrt{N}}. \tag{3.4}$$

An autocorrelation estimate is considered statistically different from zero if it exceeds approximately twice its standard error.

A second statistical estimate useful in time series analysis is the PAC estimate of coefficient j at lag k, ϕ_{kj}. The PAC are found in the following manner:

$$\rho_j = \phi_{kl}p_{j-1} + \phi_{k2}p_{j-2} + \ldots + \phi_{k(k-1)}p_{jk-1} + \phi_{kk}p_{j-k}, \quad j = 1, 2, \ldots, k$$

or

$$\begin{bmatrix} 1 & \rho_1 & \rho_2 & \cdots & \rho_{k-1} \\ \rho_1 & 1 & \rho_1 & \cdots & \rho_{k-2} \\ \vdots & \vdots & \vdots & \cdots & \vdots \\ \rho_{k-1} & \rho_{k-2} & \rho_{k-3} & \cdots & 1 \end{bmatrix} \begin{bmatrix} \phi_{k-1} \\ \phi_{k2} \\ \vdots \\ \phi_{kk} \end{bmatrix} \begin{bmatrix} \rho_1 \\ \rho_2 \\ \vdots \\ \rho_k \end{bmatrix}.$$

The PAC estimates may be found by solving the above equation systems for $k = 1, 2, 3, \ldots, k$:

$$\phi_{11} = \rho_1,$$

$$\phi_{22} = \frac{\rho_2 - \rho_1^2}{1 - \rho_1^2} = \frac{\begin{vmatrix} 1 & \rho_1 \\ \rho_2 & \rho_2 \end{vmatrix}}{\begin{vmatrix} 1 & \rho_1 \\ \rho_1 & 1 \end{vmatrix}},$$

$$\phi_{33} = \frac{\begin{vmatrix} 1 & \rho_1 & \rho_1 \\ \rho_1 & 1 & \rho_2 \\ \rho_2 & \rho_1 & \rho_3 \end{vmatrix}}{\begin{vmatrix} 1 & \rho_1 & \rho_2 \\ \rho_1 & 1 & \rho_1 \\ \rho_2 & \rho_1 & 1 \end{vmatrix}}.$$

The PAC function is estimated by expressing the current autocorrelation function estimates as a linear combination of previous orders of autocorrelation estimates:

$$\hat{r}_1 = \hat{\phi}_{k1'}{}_{j-1} + \hat{\phi}_{k2^2}{}_{j-2} + \ldots + \hat{\phi}_{k(k-1)'}{}_{j+k-1} + \hat{\phi}_{kk^2}{}_{j-k}, \quad j = 1, 2, \ldots, k.$$

The standard error of the PAC function is approximately

$$Var[\hat{\phi}_{kk}] \cong \frac{1}{N} \quad \text{and} \quad S_e[\phi_{kk}] \cong \frac{1}{\sqrt{N}}.$$

The Autoregressive and Moving Average Processes

A stochastic process, or time series, can be repeated as the output resulting from a white noise input, α_t:[3]

$$\begin{aligned}\tilde{Z}_t &= \alpha_t + \Psi_1 \alpha_{t-1} + \Psi_2 \alpha_{t-2} + \dots \\ &= \alpha_t + \sum_{j=1}^{\infty} \Psi_j a_{t-j}\end{aligned} \quad . \tag{3.5}$$

The filter weight, Ψ_j, transforms input into the output series. One normally expresses the output, $\tilde{Z}_t$, as a deviation of the time series from its mean, μ, or origin

$$\tilde{Z}_t = Z_t - \mu.$$

The general linear process leads one to represent the output of a time series, $\tilde{Z}_t$, as a function of the current and previous value of the white noise process, α_t, which may be represented as a series of shocks. The white noise process, α_t, is a series of random variables characterized by

$$\begin{aligned} E[\alpha_t] &\cong 0 \\ Var[\alpha_t] &= \sigma_\alpha^2 \\ \gamma_k = E[\alpha_t \alpha_{t+k}] &= \sigma_\alpha^2 \quad k \neq 0 \\ 0 \quad k &= 0 \end{aligned} \quad .$$

The autocorrelation function of a linear process may be given by

$$\gamma_k = \sigma_\alpha^2 \sum_{j=0}^{\infty} \Psi_j \Psi_{j+k}.$$

The backward shift operator, B, is defined as $BZ_t = Z_{t-1}$ and $B^j Z_t = Z_{t-j}$. The autocorrelation generating function may be written as

$$\gamma(B) = \sum_{k=-\infty}^{\infty} \gamma_k B^k.$$

For stationarity, the ψ weights of a linear process must satisfy that $\psi(B)$ converges on or lies within the unit circle.

[3] Please see Box and Jenkins, *Time Series Analysis*, Chap. 3, for the most complete discussion of the ARMA (p,q) models.

In an autoregressive, AR, model, the current value of the time series may be expressed as a linear combination of the previous values of the series and a random shock, α_t:

$$\tilde{Z}_t = \phi_1 \tilde{Z}_{t-1} + \phi_2 \tilde{Z}_{t-2} + \ldots + \phi_p \tilde{Z}_{t-p} + \alpha_t.$$

The autoregressive operator of order P is given by

$$\phi(B) = 1 - \phi_1 B^1 - \phi_2 B^2 - \ldots - \phi_p B^p$$

or

$$\phi(B)\tilde{Z}_t = \alpha_t. \tag{3.6}$$

In an autoregressive model, the current value of the time series, $\tilde{Z}_t$, is a function of previous values of the time series, $\tilde{Z}_{t-1}$, $\tilde{Z}_{t-2}$, ..., and is similar to a multiple regression model. An autoregressive model of order p implies that only the first p order weights are nonzero. In many economic time series, the relevant autogressive order is one. The autoregressive process of order p, AR(p) is written as

$$\tilde{Z}_t = \phi_1 \tilde{Z}_{t-1} + \alpha_t$$

or

$$(1 - \phi_1 B)\tilde{Z}_t = \alpha_t \text{ implying}$$

$$\tilde{Z}_t = \phi^{-1}(B)\alpha_t.$$

The relevant stationarity condition is $|B| < 1$ implying that $|\phi_1| < 1$. The autocorrelation function of a stationary autoregressive process

$$\tilde{Z}_t = \phi_1 \tilde{Z}_{t-1} + \phi_2 \tilde{Z}_{t-2} + \ldots + \phi_p \tilde{Z}_{t-p} + \alpha_t$$

may be expressed by the difference equation

$$P_k = \phi_1 \rho_{k-1} + \phi_2 \rho_{k-2} + \ldots + \phi_k \rho_{k-p}, \quad k>0.$$

Or expressed in terms of the Yule-Walker equation as

$$\rho_1 = \phi_1 + \phi_2 \rho_1 + \ldots + \phi_p \rho_{p-1},$$

$$\rho_2 = \phi_1 \rho_1 + \phi_2 + \ldots + \phi_p \rho_{p-2},$$

$$\bar{\bar{\rho}}_p = \phi_1\rho_{p-1} + \phi_2\rho_{p-2} + \ldots + \bar{\bar{\phi}}_p.$$

For the first-order AR process, AR(1)

$$\rho_k = \phi_1\rho_{k-1} = \bar{\bar{\phi}}_p.$$

The autocorrelation function decays exponentially to zero when ϕ_1 is positive and oscillates in sign and decays exponentially to zero when ϕ_1 is negative:

$$P_1 = \phi_1$$

and

$$\sigma_2 = \frac{\sigma_\alpha^2}{1 - \phi_1^2}.$$

The PAC function cuts off after lag one in an AR(1) process. For a second-order AR process, AR(2)

$$\tilde{Z}_t = \phi_1\tilde{Z}_{t-1} + \phi_2\tilde{Z}_{t-k} + \alpha_t$$

with roots

$$\phi(B) = 1 - \phi_1 B - \phi_2 B^2 = 0$$

and, for stationarity, roots lying outside the unit circle, ϕ_1 and ϕ_2, must obey the following conditions:

$$\phi_2 + \phi_1 < 1,$$

$$\phi_2 - \phi_1 < 1,$$

$$-1 < \phi_2 < 1.$$

The autocorrelation function of an AR(2) model is

$$\rho_k = \phi_1\rho_{k-1} + \phi_2\rho_{k-2}. \tag{3.7}$$

The autocorrelation coefficients may be expressed in terms of the Yule-Walker equations as

$$\rho_1 = \phi_1 + \phi_2\rho_2,$$

$$\rho_2 = \phi_1\rho_1 + \phi_2,$$

which implies

$$\phi_1 = \frac{\rho_1(1-\rho_2)}{1-\rho_1^2},$$
$$\phi_2 = \frac{\rho_2(1-\rho_1^2)}{1-\rho_1^2},$$

and

$$\rho_1 = \frac{\phi_1}{1-\phi_2} \quad \text{and} \quad \rho_2 = \phi_2 + \frac{\phi_1^2}{1-\phi_2}.$$

For a stationary AR(2) process,

$$-1<\phi_1<1,$$

$$-1<\rho_2<1,$$

$$\rho_1^2<\frac{1}{2}(\rho_2+1).$$

In an AR(2) process, the autocorrelation coefficients tail off after order two and the PAC function cuts off after the second order (lag).[4]

In a q-order moving average (MA) model, the current value of the series can be expressed as a linear combination of the current and previous shock variables:

$$\begin{aligned}\tilde{Z}_t &= \alpha_1 - \theta_1\alpha_{t-1} - \ldots - \alpha_q\theta_{t-q} \\ &= (1-\theta_1 B_1 - \ldots - \theta_q B_q)\alpha_t\ . \\ &= \theta(B)\alpha_t\end{aligned}$$

The autocovariance function of a q-order moving average model is

$$\gamma_k = E[(\alpha_t - \theta_1\alpha_{t-1} - \ldots - \theta_q\alpha_{t-q})(\alpha_{t-k} - \theta_1\alpha_{t-k-1} - \ldots - \theta_q\alpha_{t-k-q})].$$

[4] A stationary AR(p) process can be expressed as an infinite weighted sum of the previous shock variables

$$\tilde{Z}_t = \phi^{-1}(B)\alpha_t.$$

In an invertible time series, the current shock variable may be expressed as an infinite weighted sum of the previous values of the series

$$\theta^{-1}(B)\tilde{Z}_t = \alpha_t.$$

The autocorrelation function, ρ_k, is

$$\rho_k = \begin{cases} \dfrac{-\theta_k + \theta_1\theta_{k+1} + \ldots + \theta_{q-k}\theta_q}{1 + \theta_1^2 + \ldots + \theta_q^2} & k = 1, 2, \ldots, q \\ 0 & k>q \end{cases}.$$

The autocorrelation function of an MA(q) model cuts off, to zero, after lag q and its PAC function tails off to zero after lag q. There are no restrictions on the moving average model parameters for stationarity; however, moving average parameters must be invertible. Invertibility implies that the π weights of the linear filter transforming the input into the output series, the π weights lie outside the unit circle:

$$\pi(B) = \Psi^{-1}(B) = \sum_{j=0}^{a} \phi^j B^j.$$

In a first-order moving average model, MA(1)

$$\tilde{Z}_t = (1 - \theta_1 B)\alpha_t$$

and the invertibility condition is $|\theta_1| < 1$. The autocorrelation function of the MA (1) model is

$$\rho_k = \frac{-\theta_1}{1 + \theta_1^2} \quad k = 1,\ k>2.$$

The PAC function of an MA(1) process tails off after lag one and its autocorrelation function cuts off after lag one.

In a second-order moving average model, MA(2)

$$\tilde{Z}_t = \alpha_t - \theta_1\alpha_{t-1} - \theta_2\alpha_{t-2},$$

the invertibility conditions require

$$\theta_2<\theta_1<1,$$

$$\theta_2 - \theta_1<1,$$

$$-1<\theta_2<1.$$

The autocorrelation function of the MA(2) is

$$\rho_1 = \frac{-\theta_1(1 - \theta_2)}{1 + \theta_1^2 + \theta_1^2},$$

$$\rho_2 = \frac{-\theta_2}{1 + \theta_1^2 + \theta_1^2},$$

and

$$\rho_k = \theta \quad for\ k>3.$$

The PAC function of an MA(2) tails off after lag two.

In many economic time series, it is necessary to employ a mixed autoregressive-moving average (ARMA) model of the form

$$\tilde{Z}_t = \phi_1 \tilde{Z}_{t-1} + \ldots + \phi_p \tilde{Z}_{t-p} + \alpha_t - \theta_1 \alpha_{t-1} - \ldots - \theta_q \alpha_{t-q} \tag{3.8}$$

or

$$(1 - \phi_1 B - \phi_2 B^2 - \ldots - \theta_p B^p)\tilde{Z}_t = (1 - \theta_1 B - \theta_2 B^2 - \ldots - \theta_q B^q)\alpha_t$$

that may be more simply expressed as

$$\phi(B)\tilde{Z}_t = \theta(B)\alpha_t.$$

The autocorrelation function of the ARMA model is

$$\rho_k = \phi_1 \rho_{k-1} + \phi_2 \rho_{k-2} + \ldots + \phi_p \rho_{k-p}$$

or

$$\phi(B)\rho_k = 0.$$

The first-order autoregressive–first-order moving average operator ARMA (1,1) process is written as

$$\tilde{Z}_t - \phi_1 \tilde{Z}_{t-1} = \alpha_t - \theta_1 \alpha_{t-1}$$

or

$$(1 - \phi_1)\tilde{Z}_t = (1 - \theta 1 B)\alpha_t.$$

The stationary condition is $-1 < \phi_1 < 1$ and the invertibility condition is $-1 < \phi_1 < 1$. The first two autocorrelations of the ARMA (1,1) model are

$$\rho_1 = \frac{(1 - \phi_1 \theta_1)(\phi_1 - \theta_1)}{1 + \theta_1^2 - 2\phi_1 \theta_1}$$

and

$$\rho_2 = \phi_1 \rho_1.$$

The PAC function consists only of $\phi_{11} = \rho_1$ and has a damped exponential.

An integrated stochastic progress generates a time series if the series is made stationary by differencing (applying a time-invariant filter) the data. In an integrated process, the general form of the time series model is

$$\phi(B)(1 - B)^d X_t = \theta(B)\varepsilon_t, \tag{3.9}$$

where $\phi(B)$ and $\theta(B)$ are the autoregressive and moving average polynominals in B of orders p and q, ε_t is a white noise error term, and d is an integer representing the order of the data differencing. In economic time series, a first-difference of the data is normally performed.[5] The application of the differencing operator, d, produces a stationary ARMA(p,q) process. The autoregressive integrated moving average, ARIMA, model is characterized by orders p, d, and q [ARIMA (p,d,q)]. Many economics series follow an RWD, and an ARMA (1,1) may be written as

$$\bar{V}^d X_t = X_t - X_{t-1} = \varepsilon_t + b\varepsilon_{t-l}.$$

An examination of the autocorrelation function estimates may lead one to investigate using a first-difference model when the autocorrelation function estimates decay slowly. In an integrated process, the corr($X_t, X_{t-\tau}$) is approximately unity for small values of time, τ.

ARMA Model Identification in Practice

Time series specialists use many statistical tools to identify models; however, the sample autocorrelation and PAC function estimates are particularly useful in modeling. Univariate time series modeling normally requires larger data sets than regression and exponential smoothing models. It has been suggested that at least 30–50 observations be used to obtain reliable estimates.[6] One normally calculates the sample autocorrelation and PAC estimates for the raw time series and its first (and possibly second) differences. The failure of the autocorrelation function estimates of the raw data series to die out as large lags implies that a first difference is necessary. The autocorrelation function estimates of a MA(q) process should cut

[5] Box and Jenkins, *Time Series Analysis*. Chapter 6; C.W.J. Granger and Paul Newbold, *Forecasting Economic Time Series*. Second Edition (New York: Academic Press, 1986), pp. 109–110, 115–117, 206.

[6] Granger and Newbold, *Forecasting Economic Time Series*. pp. 185–186.

off after q. To test whether the autocorrelation estimates are statistically different from zero, one uses a t-test where the standard error of $\upsilon\tau$ is[7]

$$n^{-1/2}[1+2(\rho_1^2+\rho_2^2+\ldots+\rho_q^2)]^{1/2} \quad for \quad \tau>q.$$

The PAC function estimates of an AR(p) process cut off after lag p. A t-test is used to statistically examine whether the PAC are statistically different from zero. The standard error of the PAC estimates is approximately

$$\frac{1}{\sqrt{N}} \quad \text{for} \quad K>p.$$

One can use the normality assumption of large samples in the t-tests of the autocorrelation and PAC estimates. The identified parameters are generally considered statistically significant if the parameters exceed twice the standard errors.

The ARMA model parameters may be estimated using nonlinear least squares. The following ARMA framework-forecasts the initial parameter estimates and assumes that the shock terms are to be normally distributed:

$$\alpha_t = \tilde{W}_t - \phi_1\tilde{W}_{t-1} - \phi_2\tilde{W}_{t-2} - \ldots - \phi_p\tilde{W}_{t-p} + \theta_1\alpha_{t-1} + \ldots + \theta_q\alpha_{t-q},$$

where

$$W_t = \bar{V}^d Z_t \quad \text{and} \quad \tilde{W}_t = W_t - \mu.$$

The minimization of the sum of squared errors with respect to the autoregressive and moving average parameter estimates produces starting values for the p order AR estimates and q order MA estimates:

$$\left.\frac{\partial e_t}{-\partial\phi_j}\right|_{\beta_0} = \mu_{j,t} \text{ and } \left.\frac{\partial e_t}{-\partial\theta_i}\right|_{\beta_0} = X_{j,t}.$$

It may be appropriate to transform a series of data such that the residuals of a fitted model have a constant variance, or are normally distributed. The log transformation is such a data transformation that is often used in modeling economic time series. Box and Cox (1964) put forth a series of power transformations useful in modeling time series.[8] The data is transformed by choosing a value of λ that is

[7] Box and Jenkins, *Time Series Analysis*. pp. 173–179.

[8] G.E. Box and D.R. Cox, "An Analysis of Transformations," *Journal of the Royal Statistical Society*, B 26 (1964), 211–243.

suggested by the relationship between the series amplitude (which may be approximated by the range of subsets) and mean:[9]

$$X_t^{\lambda} = \frac{X_t^{\lambda} - 1}{\bar{X}^{\lambda-1}}, \tag{3.10}$$

where X is the geometric mean of the series. One immediately recognizes that if $\lambda = 0$, the series is a logarithmic transformation. The log transformation is appropriate when there is a positive relationship between the amplitude and mean of the series. A $\lambda = 1$ implies that the raw data should be analyzed and there is no relationship between the series range and mean subsets. One generally selects the λ that minimizes the smallest residual sum of squares, although an unusual value of λ may make the model difficult to interpret. Some authors may suggest that only values of λ of -0.5, 0, 0.5, and 1.0 be considered to ease in the model building process.[10]

Many time series, involving quarterly or monthly data, may be characterized by rather large seasonal components. The ARIMA model may be supplemented with seasonal autoregressive and moving average terms:

$$\begin{aligned}&(1-\phi_1 B-\phi_2 B^2-\ldots-\phi_p B^p)(1-\phi_{1,s}B^s-\ldots-\phi_{p,s}B^pS^s)(1-B)^d\\&\quad(1-B^s)^{ds}X_t\\&=(1-\theta_1 B-\ldots-\theta_q B^q)(1-\theta_{1,s}B^s-\ldots-\theta_{q,s}B^{q,s})\alpha_t \text{ or } \theta_p(B)\Phi_p(B^s)\\&\quad\bar{V}^d\bar{V}_x^D Z_t\\&=\theta_q(B)\theta_Q(B^s)\alpha_t.\end{aligned} \tag{3.11}$$

One recognizes seasonal components by an examination of the autocorrelation and PAC function estimates. That is, the autocorrelation and PAC function estimates should have significantly large values at lags 1 and 12 as well as smaller (but statistically significant) values at lag 13 for monthly data.[11] One seasonally differences the data (a 12th-order seasonal difference for monthly data and estimates the seasonal AR or MA parameters). An RWD model with a monthly component may be written as

$$\bar{V}\bar{V}_{12}Z_t = (1-B)(1-\theta B^{12})\alpha_t. \tag{3.12}$$

The multiplicative form of the (0,1,1) × (0,1,1)12 model has a moving average operator that may be written as

[9] G.M. Jenkins, "Practical Experience with Modeling and Forecasting Time Series," *Forecasting* (Amsterdam: North-Holland Publishing Company, 1979).

[10] Jenkins, *op. cit.*, pp. 135–138.

[11] Box and Jenkins, *Time Series Analysis*, pp. 305–308.

$$(1-\theta B)(1-\theta B^{12}) = 1-\theta B-\theta B^{12}+\theta B^{13}.$$

The RWD with the monthly seasonal adjustments is the basis of the "airline model" in honor of the analysis by Professors Box and Jenkins of total airline passengers during the 1949–1960 period.[12] The airline passenger data analysis employed the natural logarithmic transformation.

There are several tests and procedures that are available for checking the adequacy of fitted time series models. The most widely used test is the Box–Pierce test, where one examines the autocorrelation among residuals, α_t:

$$\hat{v}_k = \frac{t = \sum\limits_{k+1}^{n} \alpha_t \alpha_{t-k}}{\sum\limits_{t=1}^{n} \alpha_t^2}, \quad k = 1, 2, \ldots.$$

The test statistic, Q, should be X^2 distributed with (m–p–q) degrees of freedom:

$$Q = n \sum_{k=1}^{m} \hat{v}_k^2.$$

The Ljung–Box statistic is a variation on the Box–Pierce statistic and the Ljung–Box Q statistic tends to produce significance levels closer to the asymptotic levels than the Box–Pierce statistic for first-order moving average processes. The Ljung–Box statistic, the model adequacy check reported in the SAS system, can be written as

$$Q = n(n+2) \sum_{k=1}^{m} (n=k)^{-1} \hat{v}_k^2. \tag{3.13}$$

Residual plots are generally useful in examining model adequacy; such plots may identify outliers as we noted in the chapter. The normalized cumulative periodogram of residuals should be examined.

Granger and Newbold (1977) and McCracken (2000) use several criteria to evaluate the effectiveness of the forecasts with respect to the forecast errors. In this chapter, we use the root mean square error (RMSE) criteria. One seeks to minimize the square root of the sum of the absolute value of the forecast errors squared. That is, we calculate the absolute value of the forecast error, square the error, sum the squared errors, divided by the number of forecast periods, and take the square root of the resulting calculation. Intuitively, one seeks to minimize the forecast errors. The absolute value of the forecast errors is important because if

---

[12] Box and Jenkins, *op. cit.*

one calculated only a mean error, a 5% positive error could "cancel out" a 5% negative error. Thus, we minimize the out-of-sample forecast errors. We need a benchmark for forecast error evaluation. An accepted benchmark (Mincer and Zarnowitz 1969) for forecast evaluation is a NoCH. A forecasting model should produce a lower RMSE than the NoCH model. If several models are tested, the lowest RMSE model is preferred.

In the world of business and statistics, one often speaks of autoregressive, moving average, and RWD models, or processes, as we have just introduced.

It is well known that the majority of economic series, including real Gross National Product (GDP) in the United States, follow a RWD process, and are represented with ARIMA model with a first-order moving average operator applied to the first-difference of the data. The data is differenced to produce stationary, where a process has a (finite) mean and variance that do not change over time and the covariance between data points of two series depends upon the distance between the data points, not on the time itself. The RWD process, estimated with an ARIMA (0,1,1) model, is approximately equal to a first-order exponential smoothing model. The RWD model has been supported by the work of Nelson and Plosser (1982).

In a transfer function model, one models the dynamic relationship between the deviations of input X and output Y. One is concerned with estimating the delay between the input and output. The set of weights is often referred to as the impulse response function:

$$Y_t = V_0\tilde{X}_t + V_1\tilde{X}_{t-1} + V_2\tilde{X}_{t-2}. \tag{3.14}$$

$$= V(B)\tilde{X}_t. \tag{3.15}$$

Modeling Real GDP: An Example

GDP is the market value of all goods and services produced within a country in a given period. The expenditure approach holds that GDP is the sum of personal consumption, gross investment, government spending, and net exports (exports less imports). Let us go to a source of real-business economic and financial data. The St. Louis Federal Reserve Bank has an economic database, denoted FRED, containing some 41,000 economic series, available at no cost, via the Internet, at http://research.stlouisfed.org/fred2.

If one downloaded and graphed quarterly real (in 2005 dollars) GDP data from 1947 to 2011Q1 (April 1, 2011), one sees in Chart 1 that the postwar period has been one of great, fairly consistent growth.

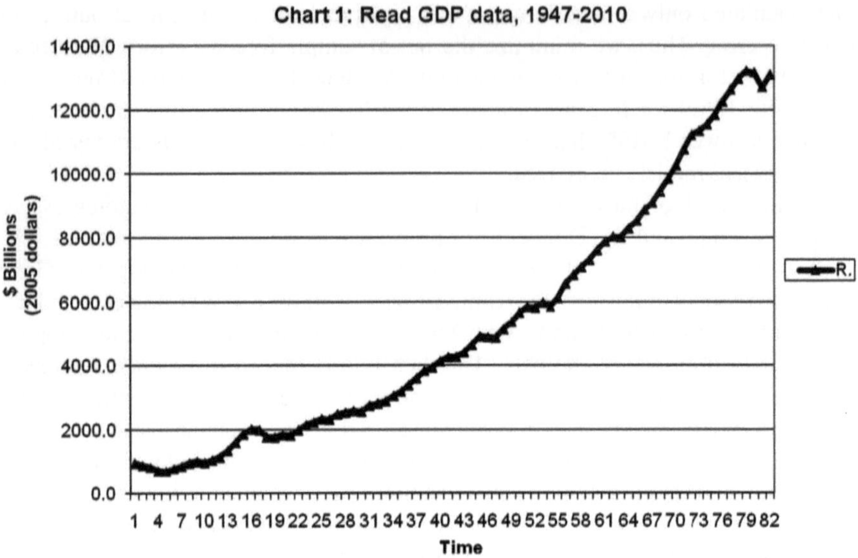

The recession of 2007–2008 is pronounced and notable, the most obvious contraction of the postwar period.

Let us examine the autocorrelation (AC) and PAC functions of the quarterly data. The raw data AC and PAC function estimates, estimated in EViews, are shown in Table 3.1, and indicate the need to (first) difference the data. One can apply the Box–Jenkins time series methodology to the real GDP data and estimate several basic models. We can take the difference of the logarithm of the series to produce stationarity and estimate a first-order autoregressive parameter to approximate the data (see Table 3.2).

We estimate an RWD model, an ARIMA (0,1,1), in Table 3.3 for the US real GDP, 1947–2011Q1. The drift term, a first-order moving average term with a 0.289 coefficient, is statistically significant, having a t-statistic of 4.89. The overall F-statistic of 31.12 indicates that the model is adequate fit. The RWD model is an adequate representation of the real GDP data generating process. One can, and should, fit other ARIMA models.[13]

The author fits an ARIMA (1,1,0) model as an additional ARIMA benchmark at the suggestion of Professor Victor Zarnowitz.[14] The ARIMA (1,1,0) has a higher F-statistics than the ARIMA (1,1,0) and a higher t-statistic on the first-order

[13] The EViews software, EViews4, in this chapter is an extremely easy system to use. The author first worked with Box–Jenkins time series model using the Nelson (1973) and Jenkins (1979) monographs and the ARIMA programs of David Pack (1982).

[14] Victor Zarnowitz was formerly emeritus of the University of Chicago, Senior Economist at TCB, and a long-term fellow Associate Editor of the author at *The International Journal of Forecasting*. Victor passed away in February 2009. Zarnowitz *magnum opus* was his 592-page *Business Cycles;* however, Victor worked within two weeks of his death in 2009. Victor published with A. Ozyildirim in 2001 in a volume dedicated to his colleague and co-author, Geoffrey Moore, a distinguished business scholar, see his 1961 volumes.

Table 3.1 Autocorrelation and partial autocorrelation function estimates of Real GDP, 1947–2011Q1

Autocorrelation	Partial correlation		AC	PAC	Q-Stat	Prob
.|********	.|********	1	0.990	0.990	256.60	0.000
.|********	.|. |	2	0.979	−0.013	508.73	0.000
.|*******|	.|. |	3	0.968	−0.013	756.34	0.000
.|*******|	.|. |	4	0.958	−0.012	999.38	0.000
.|*******|	.|. |	5	0.947	−0.005	1237.9	0.000
.|*******|	.|. |	6	0.936	−0.002	1471.9	0.000
.|*******|	.|. |	7	0.925	−0.001	1701.6	0.000
.|*******|	.|. |	8	0.915	−0.001	1926.9	0.000
.|*******|	.|. |	9	0.904	−0.003	2148.0	0.000
.|*******|	.|. |	10	0.894	−0.010	2364.8	0.000
.|*******|	.|. |	11	0.883	−0.015	2577.3	0.000
.|*******|	.|. |	12	0.871	−0.036	2785.1	0.000
.|*******|	.|. |	13	0.859	−0.037	2988.0	0.000
.|*******|	.|. |	14	0.847	−0.021	3186.0	0.000
.|****** |	.|. |	15	0.835	−0.004	3379.0	0.000
.|****** |	.|. |	16	0.822	−0.017	3567.0	0.000
.|****** |	.|. |	17	0.810	−0.008	3750.1	0.000
.|****** |	.|. |	18	0.797	−0.004	3928.2	0.000
.|****** |	.|. |	19	0.785	−0.001	4101.6	0.000
.|****** |	.|. |	20	0.772	−0.014	4270.2	0.000
.|****** |	.|. |	21	0.760	−0.006	4434.1	0.000
.|****** |	.|. |	22	0.747	−0.014	4593.2	0.000
.|****** |	.|. |	23	0.734	−0.008	4747.6	0.000
.|****** |	.|. |	24	0.722	0.007	4897.5	0.000

autoregressive parameter, 6.50. The author used the ARIMA (1,1,0) benchmark is a study of the effectiveness of TCB LEI (Guerard 2001). Both ARIMA models are adequately fit (see Table 3.4).

If one chose not to difference the real GDP data and fit a first-order autoregressive model, one finds an AR(1) parameter near 1, see Table 3.5.

The initial view of the adjusted R-square and F-statistic might lead the reader to believe that the AR(1) model was almost "truth." One must model changes in financial economic data.

Leading Economic Indicators and Real GDP Analysis: The Statistical Evidence, 1970–2002

We introduce the time series modeling process in this study because we will use TCB US composite LEI as an input to a transfer function model of US real GDP, both series being first-differenced and log-transformed. The authors test the null hypothesis that there is no statistical association between changes in the logged LEI and changes in logged real GDP in the United States. A positive and statistically

Table 3.2 Autocorrelation and partial autocorrelation function estimates of differenced Real GDP, 1947–2011Q1

Autocorrelation	Partial correlation	Lag	AC	PAC	Q-Stat	Prob
.|**** |	.|**** |	1	0.474	0.474	58.536	0.000
.|*** |	.|* |	2	0.346	0.157	89.953	0.000
.|* |	*|. |	3	0.151	−0.082	95.941	0.000
.|* |	.|. |	4	0.106	0.023	98.922	0.000
.|. |	*|. |	5	−0.016	−0.089	98.987	0.000
.|. |	.|. |	6	0.022	0.056	99.111	0.000
.|. |	.|. |	7	0.006	0.017	99.122	0.000
.|. |	.|. |	8	−0.008	−0.039	99.141	0.000
.|* |	.|* |	9	0.126	0.192	103.44	0.000
.|* |	.|. |	10	0.104	−0.011	106.39	0.000
.|. |	*|. |	11	0.044	−0.09	106.92	0.000
*|. |	*|. |	12	−0.059	−0.1	107.88	0.000
.|. |	.|. |	13	−0.005	0.062	107.88	0.000
.|. |	.|* |	14	−0.001	0.07	107.88	0.000
.|. |	.|. |	15	−0.005	−0.038	107.89	0.000
.|* |	.|* |	16	0.075	0.103	109.43	0.000
.|. |	.|. |	17	0.038	−0.033	109.82	0.000
.|. |	.|. |	18	0.058	0.001	110.76	0.000
.|* |	.|* |	19	0.096	0.071	113.34	0.000
.|* |	.|. |	20	0.092	−0.013	115.73	0.000
.|. |	.|. |	21	0.024	0.005	115.89	0.000
.|. |	.|. |	22	0.053	0.051	116.7	0.000
.|. |	.|. |	23	0.06	0.013	117.74	0.000
.|* |	.|* |	24	0.126	0.12	122.29	0.000

significant coefficient indicates that the leading indicator composite series is associated with rising real output, and leads to the rejection of the null hypothesis.

Zarnowitz (1992) examined the determinants of Real GDP, 1953–1982, using VAR models. In this analysis, we test the statistical significance of TCB LEI by adding the lags of the variable to an AR(1) model. Does the knowledge of the LEI help forecast future changes in GDP, and can past values of the GDP data predict

Table 3.3 An ARIMA RWD estimate of Real Gross Domestic Product, 1947–2011Q1

Dependent variable: DLOG(RGDP)
Method: Least squares
Date: 02/12/12, Time: 07:34
Sample(adjusted): 2 259
Included observations: 258 after adjusting endpoints
Convergence achieved after 12 iterations
Backcast: 1

Variable	Coefficient	Std. error	*t*-Statistic	Prob.
C	0.007817	0.000756	10.33377	0.0000
MA(1)	0.289085	0.059828	4.831927	0.0000
R-Squared	0.108390	Mean dependent var		0.007825
Adjusted *R*-squared	0.104907	S.D. dependent var		0.009970
S.E. of regression	0.009432	Akaike info criterion		−6.481599
Sum squared resid	0.022777	Schwarz criterion		−6.454057
Log likelihood	838.1263	*F*-Statistic		31.12102
Durbin–Watson stat	1.866243	Prob (*F*-statistic)		0.000000

Table 3.4 An ARIMA estimate of Real Gross Domestic Product, 1947–2011Q1

Dependent variable: DLOG(RGDP)
Method: Least squares
Date: 01/23/12, Time: 14:52
Sample(adjusted): 3 259
Included observations: 257 after adjusting endpoints
Convergence achieved after 3 iterations

Variable	Coefficient	Std. error	*t*-Statistic	Prob.
C	0.007875	0.000926	8.506078	0.0000
AR(1)	0.376487	0.057913	6.500889	0.0000
R-Squared	0.142170	Mean dependent var		0.007861
Adjusted *R*-squared	0.138806	S.D. dependent var		0.009972
S.E. of regression	0.009254	Akaike info criterion		−6.519720
Sum squared resid	0.021838	Schwarz criterion		−6.492100
Log likelihood	839.7840	*F*-statistic		42.26155
Durbin–Watson stat	2.067711	Prob (*F*-statistic)		0.000000

the future growth of GDP? In a recent study of univariate and time series model post-sample forecasting, Thomakos and Guerard (2004) compared RWD and transfer-function models with NoCH forecasts using rolling one-period-ahead post-sample periods. Guerard (2001) found that the AR(1) and RWD processes are adequate representations of the time series process of real GDP, given the lags of the autocorrelation and PAC functions. Guerard (2001) reported the estimated cross-correlation functions between the G7 respective LEI and real GDP for the

Table 3.5 An AR(1) estimate of Real Gross Domestic Product, 1947–2011Q1

Dependent variable: RGDP
Method: Least squares
Date: 01/23/12, Time: 08:40
Sample(adjusted): 200 259
Included observations: 60 after adjusting endpoints
Convergence achieved after 5 iterations

Variable	Coefficient	Std. error	*t*-Statistic	Prob.
C	14,253.05	885.7279	16.09191	0.0000
AR(1)	0.972952	0.009076	107.1982	0.0000
R-Squared	0.994978	Mean dependent var		11,944.42
Adjusted *R*-squared	0.994892	S.D. dependent var		1137.426
S.E. of regression	81.29588	Akaike info criterion		11.66683
Sum squared resid	383,323.1	Schwarz criterion		11.73664
Log likelihood	−348.0050	*F*-Statistic		11,491.46
Durbin–Watson stat	1.033644	Prob (*F*-statistic)		0.000000

1970–2000 period, and found that the resulting transfer function models were statistically significant in forecasting real GDP in the G7 nations.

In this chapter, the authors report the estimated autocorrelation and PAC functions of the US real GDP, 1963–March 2002, shown in Table 3.1. EViews is used in the analysis. Let us look at Table 3.6, the estimated autocorrelation PAC functions of real quarterly US GDP, March 1963–March 2002. The estimated autocorrelation function decays gradually, falling from 0.979 for a one period (quarter lag), 0.958 for a two quarter lag, to 0.584 for a 20 quarter lag, and 0.318 for a 36 quarter lag. The estimated PAC function is characterized by the "spike" at a one quarter lag. The first estimated partial autocorrelation is 0.979, and the second partial autocorrelation is −0.005. The US real GDP series can be estimated as an RWD series for the 1963–2002 period. The estimated functions substantiate the estimation of the first-order moving average operator of the first-differenced, log-transformed US real GDP series, denoted RWD, shown in Table 3.7. Guerard (2001) used an autoregressive variation of the RWD model as a forecasting benchmark. The residuals of the RWD model show few deviations from normality. The RWD is a statistically adequately fitted model. We estimate the cross-correlation function of the LEI and real GDP for an initial 32 quarter estimation period, following Thomakos and Guerard (2004), and use the 1978–March 2002 period for initial US post-sample evaluation. Similar estimations are derived for real GDP series in France (FR), Germany (GY), and the UK (see Table 3.8). The LEI are statistically significantly associated with real GDP during the 1978–2002 period, as are shown in the respective country GDP regressions in Table 3.8. The lag structures of the models were discussed in Guerard (2001), and we refer the reader to the initial modeling and forecasting analysis. The statistical significance of the transfer functions in Table 3.3 leads one to reject

Table 3.6 Correlogram of USGDP

Date: 05/09/03 Time: 14:28
Sample: 1 158
Included observations: 157

Autocorrelation Partial Correlation

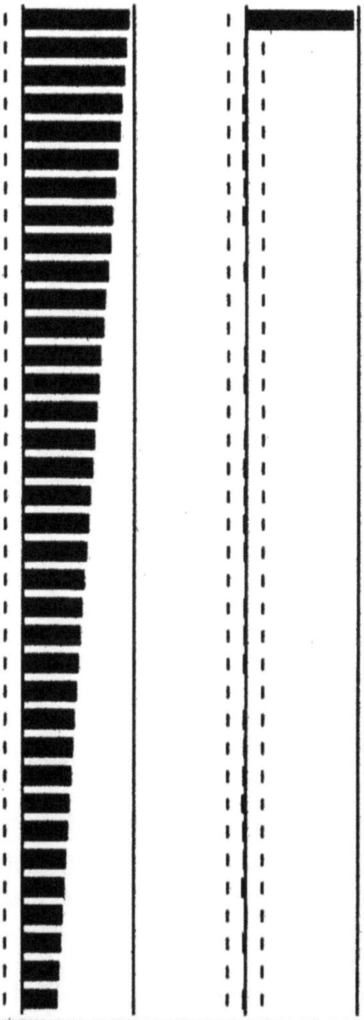

	AC	PAC	Q-Stat	Prob
1	0.979	0.979	153.31	0.000
2	0.958	-0.005	301.09	0.000
3	0.937	-0.010	443.41	0.000
4	0.916	-0.024	580.18	0.000
5	0.894	-0.024	711.33	0.000
6	0.871	-0.022	836.82	0.000
7	0.849	-0.015	956.70	0.000
8	0.826	-0.021	1071.0	0.000
9	0.804	0.003	1179.9	0.000
10	0.781	-0.011	1283.6	0.000
11	0.760	0.008	1382.4	0.000
12	0.739	0.004	1476.5	0.000
13	0.719	-0.005	1566.1	0.000
14	0.699	-0.011	1651.3	0.000
15	0.679	0.002	1732.4	0.000
16	0.660	-0.004	1809.4	0.000
17	0.640	-0.012	1882.4	0.000
18	0.621	-0.003	1951.7	0.000
19	0.602	-0.011	2017.4	0.000
20	0.584	-0.006	2079.4	0.000
21	0.566	0.004	2138.2	0.000
22	0.548	-0.001	2193.7	0.000
23	0.531	-0.004	2246.2	0.000
24	0.514	-0.013	2295.7	0.000
25	0.497	0.005	2342.4	0.000
26	0.480	-0.010	2386.4	0.000
27	0.464	-0.007	2427.7	0.000
28	0.448	-0.013	2466.5	0.000
29	0.431	-0.018	2502.7	0.000
30	0.414	-0.016	2536.4	0.000
31	0.397	-0.001	2567.7	0.000
32	0.381	-0.020	2596.6	0.000
33	0.365	0.008	2623.3	0.000
34	0.349	-0.008	2648.0	0.000
35	0.333	0.001	2670.7	0.000
36	0.318	-0.013	2691.5	0.000

Table 3.7 Random walk with drift time series model of Real US GDP

Dependent variable: DLUSGDP				
Variable	Coefficient	Std. error	*t*-Statistic	Prob.
C	0.008	0.0001	8.149	0.000
MA(1)	0.218	0.087	2.507	0.013
R-Squared	0.061			
Adjusted *R*-squared	0.053			
S.E. of regression	0.0086	Akaike info criterion		−6.6575
Sum squared resid	0.0093	Schwarz criterion		−6.6129
Log likelihood	428.08	F-statistic		8.1570
Durbin–Watson stat	1.92	Prob(F-statistic)		0.0050

Table 3.8 Post-sample regression coefficients of the leading economic indicators, 1978–March 2002

Country	Const.	LEI (−1)	LEI (−2)	LEI (−3)	LEI (−4)	AR(1)	Adjusted *R*-squared	*F*-Statistic
USA (t)	0.005	0.337	0.060	0.141		0.053	0.283	10.400
	7.200	4.800	0.890	2.130		0.480		
UK	0.005			0.214		−0.166	0.088	5.600
	7.500			2.610		−2.300		
Germany	0.004	0.242		0.211		−0.250	0.102	4.610
	5.750	2.610		2.370		−2.300		
France	0.004		0.140	0.133	−0.064	0.038	0.058	2.470
	7.960		1.930	1.870	−0.910	0.360		
Japan	0.005	0.217				−0.437	0.174	11.030
	5.860	2.900				−4.660		
Canada	0.008		0.306	0.036	−0.263	0.150	0.240	3.290
	4.880		2.340	0.270	−2.100	0.640		
Italy	0.004		0.132	−0.089	−0.009	−0.050	0.059	1.460
	4.670		2.260	−1.480	−1.490	−0.240		

the null hypothesis of no statistical association changes in the LEI and changes in real GDP. The statistically significant lags in the cross-correlation functions show how past values of the LEI series are associated with the current values of the respective real GDP. That is, the LEI series lead their respective real GDP series and can be used as inputs to transfer function models of real GDP. The multiple regressions of the post-sample period are generally statistically significant at the 1% level, as shown by their respective *F*-statistics of the regressions. The exception to this result is the French real GDP estimate, see Table 3.8, that is significant at approximately the 5% level. Thus, the estimation of the transfer function is statistically significant relative to simply using an AR(1) time series model.

US and G7 Post-sample Real GDP Forecasting Analysis

In this section, the author estimates several time series models for the US leading indicators and Real GDP, and corresponding models for the G7 nations. A simple autoregressive variation on the random walk model, an ARIMA (1,1,0), is estimated to serve as a naïve, forecasting model. The ARIMA model is referred to as the RWD Model. The transfer function model uses the LEI series as the input to the Real GDP (output) series. We will evaluate the forecasting performances of the models with respect to their RMSE, defined as the square root of the sum of the individual observation forecast errors squared. The most accurate forecast will have the smallest forecast error squared and hence the smallest RMSE. The RMSE criteria are proportional to the average squared error criteria used in Granger and Newbold (1977). One can estimate models using 32 quarters of data and forecast one-step-ahead. We compare the forecasting accuracy of four models of the US real GDP. The models tested are (1) the transfer function model in which TCB composite index of lLEI is lagged three quarters, denoted TF; (2) a NoCH forecast; (3) the simple RWD model; and (4) a simple transfer function model in which TCB composite index of LEI is lagged one period, denoted TF1. One finds that the three-quarter of lagged LEI transfer function is the most accurate out-of-sample forecasting model for the US real GDP, although there is no statistically significant differences in the rolling one-period-ahead root mean square forecasting errors of the RWD, TF, and TF1 models.

The one-period-ahead quarterly RMSE for the 1978–March 2002 period of Real GDP are shown in Table 3.9. Thus, the US leading indicators lead Real GDP, as one should expect, and the transfer function model produces lower forecast errors than the univariate model, and a naive benchmark, the NoCH model. The reader notes that the transfer function model uses a one-quarter lag that produces forecasts that are not statistically different from the three-quarter lags suggested from the estimated cross-correlation function.

The model forecast errors are not statistically different (the *t*-value of the paired differences of the univariate and TF models is 0.91). An analysis of the rolling one-period-ahead RMSE produces somewhat different results for post-sample modeling than the use of one long period of post-sample period. The multiple regression models indicate statistical significance in the US composite index of LEI for the 1978–March 2002 period. One does not find that the transfer function model forecast errors are (statistically) significantly lower than univariate ARIMA model (RWD) errors in a rolling one-period-ahead analysis. The authors prefer to measure forecasting performance in the rolling period manner (as we often live in a one-period-ahead forecasting regime).

The RMSE of the G7 nations cast doubt as to the effectiveness of the LEI as a statistically significant input in transfer function models forecasting real GDP. Transfer function model forecasts of real GDP, using TCB do not significantly reduce RMSE relative to the RWD model forecasts during the 1978–March 2002 period. Please see Table 3.10.

Table 3.9 Post-sample accuracy of the US Real GDP models using The Conference Board LEI in the transfer function model

Model	RMSE
No-change	0.0117
RWD	0.0086
TF1	0.0080
TF	0.0079

Table 3.10 Post-sample accuracy of Real GDP models using TCB LEIs in the transfer function model

Nation	Model	Input source	RMSE
GR	NoCH		0.0114
	RWD		0.0109
	TF	TCB	0.0106
FR	NoCH		0.0081
	RWD		0.0065
	TF	TCB	0.0070
JP	NoCH		0.0177
	RWD		0.0152
	TF	TCB	0.0163
UK	NoCH		0.0106
	RWD		0.0090
	TF	TCB	0.0089

Table 3.11 Post-sample root mean square errors of real US GDP, 1982–2002

Estimation modeling periods	RMSE
32	5.31
36	5.18
40	5.19
44	4.99
48	4.99
52	5.03
56	5.05
60	5.08
NoCh	8.09

One may ask why 32 observations were used. Why not use 60 observations of past real GDP to estimate the models? If one sought to minimize the forecasting error from 1982 to June 2002, and one varied the estimation modeling periods, one finds that the 32-quarter estimation is quite reasonable, see Table 3.11. The 40- and 44-quarter estimation periods produce the lowest real RMSE, although the differences are not statistically significant.

Summary

This chapter examined the predictive information in TCB LEI for the United States, the UK, Japan, and France. We find that TCB LEI and FIBER short-term LEI are statistically significant in modeling the respective real GDP changes during the 1970–2000 period. One rejects the null hypothesis of no association between changes in LEI and changes in real GDP in the United States, and the G7 nations. If one uses a rolling 32-quarter estimation period and a one-period-ahead forecasting RMSE calculation, the LEI forecasting errors are not significantly lower than the univariate ARIMA model forecasts. In Chap. 6, we estimate additional time series models and introduce the reader to causality testing.

References

Box, G.E.P. and D.R. Cox. 1964. "An Analysis of Transformations." *Journal of the Royal Statistical Society* B 26, 211–243.

Box, G.E.P. and G.M. Jenkins. 1970. *Time Series Analysis: Forecasting and Control*. San Francisco: Holden-Day.

Burns, A. F. and W.C. Mitchell. 1946. *Measuring Business Cycles*. New York, NBER.

Fisher, I. 1911. *The Purchasing Power of Money*, New York: Macmillan.

Friedman, M. and A. Schwartz. 1963. "Money and Business Cycles." *Review of Economics and Statistics* 45, pp. 32–64.

Granger, C.W.J. and P. Newbold. 1977. *Forecasting Economic Time Series*. New York Academic Press.

Guerard, J. B., Jr. 2004. "The Forecasting Effectiveness of the U.S. Leading Economic Indicators: Further Evidence and Initial G7 Results." In P. Dua, Ed., *Business Cycles and Economic Growth: An Analysis Using Leading Indicators*. New York: Oxford University Press, 174–187.

Guerard, J. B., Jr. 2001. "A Note on the Forecasting Effectiveness of the U.S. Leading Economic Indicators." *Indian Economic Review* 36, pp. 251–268.

Hamilton, J.D. 1994. *Time Series Analysis*, Princeton, Princeton University Press.

Jenkins, G.M. 1979. *Practical Experiences with Modelling and Forecasting Time Series*. Jersey, Channel Islands: a GJP Publication.

Ljung, G.M. and G.E.P. Box. 1978. "On a measure of lack of fit in time series models." *Biometrika* 65, 297–303.

McCracken, M. 2000. "Robust out-of-sample inference." *Journal of Econometrics 39*, 195–223.

McLeod, A.I. and W.K. Li. 1983. "Diagnostic checking ARMA time series models using squared residual autocorrelations." *Journal of Time Series Analysis* 4, 269–273.

Mitchell, W.C. 1913. *Business Cycles*. New York, Burt Franklin reprint.

Mitchell, W.C. 1951. *What Happens During Business Cycles: A Progress Report*. New York, NBER.

Moore, G.H. 1961. *Business Cycle Indicators*. 2 volumes, Princeton, Princeton University Press.

Montgomery, A.L., V. Zarnowitz, R.S. Tsay, and G.C. Tiao. 1998. "Forecasting the U.S. Unemployment Rate," *Journal of the American Statistical Association* 93, 478-493.

Nelson, C.R. and C.I. Plosser. 1982. "Trends and Random Walks in Macroeconomic Time Series." *Journal of Monetary Economics* 10, 139–162.

Nelson, C. R. 1973. *Applied Time Series Analysis for Managerial Forecasts*. San Francisco: Holden-Day. Chapters 5 and 6.

Pack, D. J. 1982. "A Practical Overview of ARIMA Models for Time Series Forecasting", in S. Makridakis and S. C. Wheelwright, Eds. *The Handbook of Forecasting: A Manager's Guide*. New York: John Wiley & Sons.
Theil, H. 1966. *Applied Economic Forecasting*. Amsterdam, North-Holland.
Thomakos, D. and J. Guerard. 2004. "Naïve, ARIMA, Transfer Function, and VAR Models: A Comparison of Forecasting Performance." *The International Journal of Forecasting 20*, 53–67.
West, K. and M. McCracken. 1998. "Regression-based Tests of Predictive Ability." *International Economic Review* 39, 817–840.
Zarnowitz, V. 2004. "The Autonomy of Recent US Growth and Business Cycles". In P. Dua, Ed., *Business Cycles and Economic Growth: An Analysis Using Leading Indicators.* New York: Oxford University Press, 44–82.
Zarnowitz, V. 2001. "The Old and the New in the U.S. Economic Expansion." The Conference Board. EPWP #01–01.
Zarnowitz, V. and A. Ozyildirim. 2001. "On the Measurement of Business Cycles and Growth Cycles." *Indian Economic Review* 36, 34–54.
Zarnowitz, V. 1992. *Business Cycles: Theory, History, Indicators, and Forecasting*. Chicago, University of Chicago Press.

Chapter 4
Regression Analysis and Multicollinearity: Two Case Studies

In this chapter, we explore two applications of regression modeling: the question of regression-weighting of GNP forecasts and the issue of estimating models associated with security totals returns. We examine the forecasting of GNP by major econometric firms and the modeling of security returns as a function of well-known investment variables and strategies. We illustrate regression analysis and problems with highly correlated independent variables. We will refer to the problem of high correlation among independent variables as multicollinearity.

The first case study involves combining econometric services' forecasts of GNP. In combining economic forecasts a problem often faced is that the individual forecasts display some degree of dependence. We discuss latent root regression (LRR) for combining collinear GNP forecasts. Guerard and Clemen (1989) results indicate that LRR produces more efficient combining weight estimates (regression parameter estimates) than ordinary least squares estimation (OLS), although out-of-sample forecasting performance is comparable to OLS. Researchers appear to have reached agreement, or consensus, regarding the value of combining forecasts. Performance, measured in terms of a variety of error summary statistics, can be improved by combining multiple forecasts. There is an extensive literature on combining forecasts that can be traced back to Bates and Granger (1969), reached a peak with Winkler and Makridakis (1983), Clemen and Winkler (1986), and Granger (1989), and was documented in a bibliography by Clemen (1989). An important unanswered question, however, regards what combination procedure to use.

There are many ways of determining these weights, and the aim was to choose a method which was likely to yield low errors for the combined forecasts. Bates and Granger, denoted as BG in many Granger references, (1969) assumed that the performance of the individual forecasts would be consistent over time in the sense that the variance of errors for the two forecasts could be denoted by ${\sigma_1}^2$ and ${\sigma_2}^2$ for all values of time, t. It was further assumed that both forecasts would be unbiased (either naturally or after a correction had been applied). The combined forecast would be obtained by a linear combination of the two sets of forecasts,

J.B. Guerard, Jr., *Introduction to Financial Forecasting in Investment Analysis*,
DOI 10.1007/978-1-4614-5239-3_4,

giving a weight k to the first set of forecasts and a weight $(1 - k)$ to the second set, thus making the combined forecast unbiased. The variance of errors in the combined forecast, σ_c^2, can then be written:

$$\sigma_c^2 = k^2\sigma_1^2 + (1-k)^2\sigma_2^2 + 2\rho k\,\sigma_1(1-k)\sigma_2, \tag{4.1}$$

where k is the proportionate weight given to the first set of forecasts and ρ is the correlation coefficient between the errors in the first set of forecasts and those in the second set. The choice of k should be made so that errors of the combined forecasts are small: more specifically, we chose to minimize the overall variance, σ_c^2. Differentiating with respect to k, and equating to zero, we get the minimum of σ_c^2, occurring when

$$k = \frac{\sigma_2^2 - \rho\sigma_1\sigma_2}{\sigma_1^2 - \sigma_2^2 - 2\rho\sigma_1\sigma_2}. \tag{4.2}$$

In the case where $\rho = 0$, this reduces to

$$k = \sigma_2^2/\left(\sigma_1^2 + \sigma_2^2\right). \tag{4.3}$$

It can be shown that if k is determined by (4.1), the value of σ_c^2 is no greater than the smaller of the two *individual* variances.[1]

The optimum value for k is not known at the commencement of combining forecasts. The value given to the weight k would change as evidence was accumulated about the relative performance of the two original forecasts. Thus the combined forecast for time period T, C_T, is more correctly written as

$$C_T = k_T f_{1,T} + (1-k_T) f_{2,T}, \tag{4.4}$$

where $f_{1,T}$ is the forecast at time T from the first set and $f_{2,T}$ is the forecast at time T from the second set.

Thought should be given to the possibility that the performance of one of the forecasts might be changing over time (perhaps improving) and that a method based on an estimate of the error variance since the beginning of the forecast might not therefore be appropriate.

Granger (1989) defined good forecasting methods (defined by us as those which yield low mean-square forecast error) are likely to possess properties such as:

(a) The average weight k should approach the optimum value, defined by (2), as the number of forecasts increased—provided that the performance of the forecasts is stationary.

---

[1] The reader will see a variation of (4.1) and (4.2) in Chap. 5 when we discuss optimal security weights in a portfolio. The Bates and Granger optimal forecast weighting is a variation of the optimal Markowitz (1959) two-asset security calculation.

(b) The weights should adapt quickly to new values if there is a lasting change in the success of one of the forecasts.
(c) The weights should vary marginally from the optimum value.

This last point is included since property (a) is not sufficient on its own.[2] In addition to these properties, there has been an attempt to restrict methods to those which are moderately simple, in order that they can be of use to businessmen.

Model building can be tested in combining forecasts. If we had available all the information, the so-called perfect foresight answer, upon which all the forecasts are based, then we would build the complete model. There would be no need for out-of-sample or post-sample forecasting periods. In most cases, only the individual forecasts are available, rather than the information they are based on, and so combining is appropriate. In the BG combinations these data were not used efficiently. For example, if $f_{n,1}$, $g_{n,1}$ are a pair of one-step forecasts of y_{n+1}, made at time n, and if the y_t series as stationary, then the unconditional mean

$$m_n = \frac{1}{n}\sum_{j=1}^{n} y_{t-j} \tag{4.5}$$

is also a forecast of y_{n+1} available at time n, although usually a very inefficient one. This new forecast can be included in the combination, giving

$$c_{n+1} = \alpha_1 m_n + \alpha_2 f_{n,1} + \alpha_3 g_{n,1} \tag{4.6}$$

as the combined forecast. The weights α_j can be obtained by regressing $c_{n,1}$ on y_{n+1} as discussed in Granger and Ramanathan (1984). Whether the weights α_j should add to one depends on whether the forecasts are unbiased and the combination is required to be unbiased. Before combining, it is usually a good idea to unbias the component forecasts. Thus, if $w_{w,1}$ is a set of one-step forecasts, run a regression

$$y_{n+1} = a + bw_{n,1} + \varepsilon_{n+1} \tag{4.7}$$

and check whether $a = 0$, $b = 1$, and if ε_n is white noise. If any of these conditions do not hold, an immediately apparently superior forecast can be achieved and these should be used in any combination.

In all these extensions of the original combining technique, combinations have been linear, only single-step horizons are considered, and the data available have been assumed to be just the various forecasts and the past data of the series being forecast. On this last point, it is clear that other data can be introduced to produce further forecasts to add to the combinations, or Bayesian techniques could

---

[2] Granger (1989) additionally pointed out that if the optimum value for k is 0.3, one may still obtain poor combined forecast if k takes two values only, being 0 on 60% of occasions and 1.0 on the remaining 40%.

be used to help determine the weights. The fact that only linear combinations were being used was viewed as an unnecessary restriction from the earliest days, but sensible ways to remove this estimation were unclear.

Procedures suggested by Bates and Granger (1969), with subsequent extensions and applications by Newbold and Granger (1974) and Winkler (1981) among others, model the forecast errors with a multinormal process, the parameters of which determine the combining weights. A number of alternative combining procedures have also been proposed, including simple averages (Makridakis and Winkler 1983), unrestricted regressions (Granger and Ramanathan 1984), weighting procedures based on assessments of which forecast might perform best (Bunn 1975; Clemen and Guerard 1989), and various ad hoc procedures (Ashton and Ashton 1985). The basic question is whether equally weighted composite forecasting models outperform statistically based forecast models.

In developing composite models using the multinormal model or related regression approaches one major problem is that the covariance matrix must typically be estimated with relatively small quantities of data. This results in unstable estimation of the covariance matrix and even more unstable estimation of the combining weights (Kang 1986). Furthermore, for economic forecasting the problem is exacerbated by the fact that different forecaster errors are typically highly correlated; correlations above 0.8 are not at all unusual (Clemen and Winkler 1986; Figlewski and Urich 1983).

We explore the possibility of using LRR (Webster et al. 1974; Gunst et al. 1976) as a procedure for combining dependent forecasts. This approach provides an explicit framework for analysis of collinear data through the mathematics of latent roots and vectors. The data we analyze (GNP forecasts studied in Clemen and Winkler 1986) display pairwise correlations of forecast errors between 0.82 and 0.96. Given these relatively high correlations as well as Kang's demonstration of the instability of the estimated weights in this data set, it seems reasonable to think that LRR might improve on the performance of OLS.

We assume that at time $t - 1$ we have access to k forecasts, $f_t = (f_{1t}, \ldots, f_{kt})$, for θ_t. We can write θ_t stochastically in terms of the (possibly biased) forecasts f_{it},:

$$\theta_t = a_i + b_i f_{it} + u_{it}, \tag{4.8}$$

where each $u_t = (u_{lt}, \ldots, u_{kt})'$ is an independent realization from a normal process with mean vector $(0, \ldots, 0)'$ and covariance matrix $\sum$. At time $t - 1$, we have available past observations (forecasts and actual values) for time $t = 1, \ldots, t - 1$. To represent these data we will adopt the following notation:

$$[\theta, F] = \begin{pmatrix} \theta_1 & 1 & f_{1,1} & \cdots & f_{k,1} \\ \cdot & \cdot & \cdot & & \cdot \\ \theta_{t-1} & 1 & f_{1,t-1} & \cdots & f_{k,t-1} \end{pmatrix}. \tag{4.9}$$

We include the vector of ones because, in general, we will be estimating regression coefficients including a constant term.

Multiply each of the different equations (4.9) by a factor γ_i such that $\sum \gamma i = 1$. Then combine equation (4.1) to obtain the following regression representation:

$$\begin{aligned}\theta_t &= \sum \gamma_i a_i + \sum \gamma_i b_i f_{it} + \sum \gamma_i \mu_{it} \\ &= \beta_0 + \beta_1 f_{1t} + \ldots + \beta_k f_{kt} + \varepsilon_t \\ &= f_t{}^* \beta + \varepsilon_t,\end{aligned} \tag{4.10}$$

where

$$\begin{aligned}\beta &= (\beta_0, \ldots, \beta_k)' = \left(\sum \gamma_i a_i, \gamma_1 b_1, \ldots, \gamma_k b_k\right)' \\ f_t{}^* \beta &= (1, f_{1t}, \ldots, f_{kt})\end{aligned}$$

and

$$\varepsilon_t = \sum \gamma_i \mu_{it}.$$

The distributional assumptions regarding μ_t imply that the regression equation error terms ε_t obey standard OLS assumptions. Therefore, the OLS estimator of β is given by the familiar expression

$$\beta^* = (F'F)^{-1} F' \theta. \tag{4.11}$$

As usual, β^* is the best linear unbiased estimator of β, and, assuming stationarity of the process through time, the forecast $\theta_t{}^* = f_t{}^* \beta^*$ is the best linear unbiased predictor of θ_t.

In the event of multicollinearity in the F matrix, β^* (and hence $\theta_t{}^*$) can be inefficient. If the process is stationary, one solution to the problem of multicollinear regressors is simply to acquire more data to improve the efficiency of the estimation, thereby improving prediction performance. However, this is often not possible, especially when working with economic data. Thus, there is some motivation to consider biased estimation and prediction if the biased approach might yield a substantial improvement in terms of estimation efficiency. LRR is one such technique. The following is a brief description of the procedure, abstracted from Webster, Gunst, and Mason (1974) and Gunst, Webster, and Mason (1976). We direct the interested reader to those papers for more details.

LRR seeks to identify near-singularities in the explanatory variables and to determine their predictive value. The procedure uses this information to estimate the regression parameters β by adjusting for non-predictive near-singularities. Define the matrix A to be $n \times (k + 1)$ data matrix containing standardized-dependent and -independent variables. The correlation matrix $(A' A)$ has latent roots λ_i and corresponding latent vectors α_i defined by

$$|A'A - \lambda_i I| = 0$$

and

$$(A'A - \lambda_i I)\alpha_i = 0.$$

Denote the elements of α_i by

$$\alpha_i' = (\alpha_{0i}, \alpha_{1i}, , \ldots, \alpha_{ki})$$

and

$$\alpha_i^{0'} = (\alpha_{1i}, \ldots, \alpha_{ki}).$$

That is, α_i^0 contains all of the elements of α_i except tile first one. Also, define

$$\eta^2 = \Sigma(\theta_i - \theta)^2.$$

The OLS estimator β^* can be written as

$$\beta^* = -\eta \Sigma c_i \alpha_i{}^0,$$

where

$$c_i = \alpha_{0i} \lambda_i{}^{-1} \left(\Sigma \alpha_0^2 / \lambda_j\right)^{-1}. \tag{4.12}$$

Values of λ_i and α_{0i} close to zero indicate a non-predictive near-singularity. As α_{0i} becomes close to zero, c_i should also be close to zero. However, since λ_i is also small, c_i may be quite large, and may have a dominant effect in the estimate β^*. Gunst et al. (1976) suggest setting $c_i = 0$ for $|\lambda_l| \leq 0.3$ and $|\alpha_{0i}| \leq 0.1$, thus obtaining the LRR estimate of the parameter β. Webster et al. (1974) and Gunst et al. (1976) provide detailed geometrical interpretations and discussion of this technique.

The First Example: Combining GNP Forecasts

Clemen and Winkler (1986) studied the forecasting efficiency of Gross National Product (GN) forecasting services in the mid-1980s, using data from the fourth quarterly of 1970 to the fourth quarter of 1983. Wharton Econometrics (Wharton), Chase Econometrics (Chase), Data Resources, Inc. (DRI), and the Bureau of

Economic Analysis (BEA) made quarterly forecasts of many economic variables. Clemen and Winkler (1986) used level forecasts of nominal GNP (1970–1983), obtained directly from Wharton and BEA and from the *Statistical Bulletin* published by the Conference Board for Chase and DRI to construct growth rate forecasts (in percentage terms), and calculated the deviations from actual growth as determined from GNP reported in *Business Conditions Digest*. Forecasts with four different horizons (one, two, three, and four quarters) were analyzed. For example, the four-quarter GNP forecast predicts the percentage change for the 3-month period four quarters in the future (counting the current one). Finally, the data are divided into two periods, one for estimation and one for forecast evaluation. The estimation period runs through 1979 for each horizon, with the remaining data kept in reserve as an independent sample for forecast evaluation. For analysis of the individual forecasts, the reader is referred to Clemen and Winkler (1986) and Clemen (1986).

Clemen and Guerard (1989) tested LRR as a combining technique because of the high pairwise correlations among the individual forecasts and the instability of the estimated weights, noted by Kang (1986). However, while these observations suggest multicollinearity, we have no clear indication of the severity of the problem. Belsley, Kuh, and Welsch (1980) and Belsley (1982, 1984) have discussed diagnostics for explicit measurement of the severity of multicollinearity. We calculated variance inflation factors, condition indexes, and variance-decomposition proportions for each of the four forecast horizons. These diagnostics are reported in Table 4.1. For condition numbers (defined as the largest of the condition indexes), the value 30 is suggested as a screen; situations with larger values are then examined more closely. All our condition numbers are between 20 and 30; thus, on the basis of this diagnostic alone our data do not appear to display severe multicollinearity. For variance inflation factors (VIFs), Montgomery and Peck (1982) suggest that values from 5 to 10 indicate severe multicollinearity. Our VIFs range up to 4.6. Variance-decomposition proportions can also be used to detect multicollinearity, which is indicated by two numbers exceeding 0.5 in any one row of the variance-decomposition table. For our forecasts, the variance-decomposition calculations reveal collinearity between (1) the DRI and BEA forecasts in the one- and two-quarter horizons, (2) the Wharton and BEA forecasts in the three-quarter one, and (3) the Chase and DRI as well as the constant and BEA variables in the four-quarter horizon.[3]

To some extent, the use of these diagnostics is problematic. For instance, condition indexes are based on eigenvalues (latent roots) of the sample covariance matrix, and it is unclear to what extent models built and estimated on the basis of this diagnostic might be sensitive for relatively small sample sizes. The presence

[3] This research was supported in part by the National Science Foundation under Grant IST 8600788. We thank George Jaszi of the BEA and Donald Straszheim of Wharton, who graciously provided the forecasts from their respective econometric models. The authors are indebted to Professors S. Sharma and W.L. James for providing access to their LRR procedure as described in Sharma and James (1981). The original data for this analysis is not available from either John Guerard or Bob Clemen.

Table 4.1 Multicollinearity diagnostics for GNP forecasts

		Variance-decomposition proportions				
Horizon	Condition indexes	Constant	Wharton	Chase	DRI	BEA
1	9.78	0.68	0.00	0.03	0.02	0.07
	15.80	0.04	0.01	0.00	0.56	0.63
	17.65	0.01	0.03	0.73	0.30	0.30
	20.93	0.27	0.96	0.24	0.11	0.00
	VIF		3.38	3.86	3.25	3.24
2	11.06	0.55	0.25	0.14	0.00	0.01
	12.58	0.17	0.60	0.13	0.01	0.13
	14.06	0.16	0.11	0.62	0.01	0.23
	27.41	0.11	0.04	0.11	0.98	0.63
	VIF		1.85	2.25	4.60	3.23
3	10.94	0.71	0.00	0.26	0.01	0.00
	13.69	0.23	0.42	0.44	0.01	0.01
	18.98	0.06	0.50	0.27	0.09	0.53
	22.24	0.00	0.08	0.03	0.88	0.46
	VIF		2.54	2.40	3.93	3.27
4	7.36	0.05	0.84	0.01	0.00	0.00
	11.14	0.29	0.06	0.29	0.09	0.01
	16.39	0.01	0.03	0.62	0.84	0.00
	22.86	0.65	0.06	0.07	0.07	0.98
	VIF		1.52	2.31	2.54	2.22

of a condition index greater than 30 may be a reliable indicator of multicollinearity; however, values slightly less than 30 do not necessarily mean that effects due to multicollinearity will be unnoticeable. With regard to the variance-decomposition proportions, the Guerard and Clemen (1989) results indicated that the one-quarter DRI and BEA forecasts appear to be associated with an ill-conditioned covariance matrix. That is, the correlation coefficient between the one-quarter DRI and BEA (0.82, reported in Clemen and Winkler 1986) is the least of the pairwise correlations for this horizon. Likewise, the correlation between Wharton and BEA errors in the two-quarter analysis (0.94) is the second-lowest of the reported pairwise correlations. Given these observations, it seems reasonable to conclude that multicollinearity, perhaps at a relatively low level, was present in the Guerard and Clemen (1989) data.

Application of LRR, using the Gunst, Webster, and Mason (1976) criteria for vector deletion, produced the results shown in Table 4.2. Details regarding the latent roots and vectors and the vector deletion patterns for each analysis are available from the authors. The coefficient estimates for the Chase and DRI forecasts are highly significant in the one-quarter horizon. In the two-quarter horizon, coefficient estimates for DRI and BEA are significant, as is the DRI coefficient estimate in the three-quarter horizon.

For comparison, OLS results are also included in Table 4.2. Generally speaking, LRR and OLS produced coefficient estimates that are comparable in terms of signs

Table 4.2 LRR and OLS regression results

Horizon		Constant	Wharton	Chase	DRI	BEA	R^2
1	LRR	1.30	−0.23	0.96	0.37	−0.11	0.40
			(−0.58)	(2.83)[a]	(4.93)a	(−1.78)	
	OLS	2.18	−0.53	0.65	0.33	0.48	0.46
			(−1.28)	(1.66)	(0.92)	(1.43)	
2	LRR	1.71	0.08	−0.25	0.41	0.63	0.24
			(0.24)	(−0.69)	(2.52)[a]	(2.31)[a]	
	OLS	1.48	0.06	−0.28	0.59	0.52	0.24
			(0.20)	(−0.76)	(0.87)	(1.10)	
3	LRR	4.17	0.16	−0.62	0.32	0.76	0.18
			(0.39)	(−1.60)	(2.56)[a]	(1.56)	
	OLS	4.17	0.21	−0.59	0.20	0.82	0.18
			(0.46)	(−1.53)	(0.34)	(1.47)	
4	LRR	8.69	−0.09	−0.60	0.96	−0.08	0.12
			(−0.40)	(−1.69)	(2.03)	(−0.36)	
	OLS	10.92	−0.06	−0.47	1.10	−0.63	0.17
			(−0.28)	(−1.10)	(2.39)[a]	(−0.96)	

Values in parentheses are *t*-statistics
[a]Significance at the 0.05 level

Table 4.3 Performance of combining methods for the post-estimation evaluation period shown

Horizon	Evaluation period	Equal weights	OLS	LRR
1	80.1–82.2	2.47	2.89	2.76
2	80.1–82.3	3.60	4.19	4.40
3	80.1–82.4	4.35	4.58	4.49
4	80.1–83.1	4.45	3.67	3.71

Performance is mean absolute relative error, where absolute relative error is defined as |(actual-forecast)/actual|

and relative sizes. (While this comparison is a matter of degree, two exceptions are BEA in the one- and four-quarter horizons). On the other hand, LRR generally yielded more efficient estimates of the parameters than OLS, as measured by the *t*-statistics.

The true test of a forecasting procedure is how well it performs outside of the fitting data. Table 4.3 presents the results obtained by using the estimated models to predict actual nominal GNP for the evaluation periods shown. Guerard and Clemen (1989) included the arithmetic average (equal weights) as one of the combining procedures for use as a benchmark. The performance measure we used, mean absolute relative error, is mean absolute percentage error (MAPE) divided by 100. MAPE is a widely used forecast performance measure that allows performance comparisons among different forecast situations (see Armstrong 1985). The results in Table 4.3 show that OLS and LLR performed comparably. Given the similar estimates of the combining weights in the two analyses, this result is not surprising. The equal weights combination outperformed the regression model in all but the four-quarter horizon.

The Guerard and Clemen (1989) empirical results show that LRR produced more efficient parameter estimates than OLS. However, the similar out-of-sample performance of the two methods leads us to be somewhat ambivalent. In theory, LRR's more efficient estimation of parameters should result in more efficient predictors and hence better out-of-sample prediction performance. In light of the data's high correlations, Kang's results, and Clemen's and Winkler's (1986) results from combining these GNP forecasts using a Bayesian model, Guerard and Clemen (1989) concluded that the comparable performance of LRR and OLS is troubling. Compared to OLS, Clemen's and Winkler's Bayesian model resulted in forecasting performance improvements of about 16% in terms of mean squared error. One possible interpretation might be that Clemen's and Winkler's model, being mathematically similar to ridge regression (Lindley and Smith 1972; Hocking 1976), tended to counteract the dependence among the forecasts. Of course, other techniques are available for use with collinear data, notably principal components regression (Gunst et al. 1976) and LRR. The Guerard and Clemen (1989) motivation for trying LRR was that it differs fundamentally from ridge regression (and the related Clemen/Winkler model) in the way multicollinearity is handled. Where ridge regression depends on the estimation of a biasing parameter, principal components regression and LRR are estimated by the elimination of non-predictive near-singularities as described above. However, the Guerard and Clemen (1989) GNP forecasts appeared to be collinear enough to cause some difficulty in the OLS analysis, but not severe enough for LRR to dominate OLS.

The Second Example: Modeling the Returns of the US Equities

Our second example will address the estimations of the determinants of the US equity security monthly returns. In 1990, Harry Markowitz became the Head of the Global Portfolio Research Department (GPRD) at Daiwa Securities Trust. Harry Markowitz wanted to build stock selection models to identify mis-priced stocks, professor Markowity was familiar with the low price-earnings P/E theories of Graham and Dodd (1934) and Williams (1938). That is, low P/E stocks art performed high P/E stocks in subsequent periods. His department used fundamental data to create models for Japanese and the US securities and the researchers tested single variable and regression-weighted composite model strategies for Japan and the USA over 1974–1990. The GPRD analysis builds upon Guerard and Takano (1991) and Guerard (1990) framework. Fundamental variables, such as earnings, book value, cash flow, and sales, standard variables of financial analysis since the days of Graham and Dodd (1934). The low "P/E" work of Basu (1977) has been substantiated in the practitioner literature, see Dremen (1998). The book-to-price variable is substantiated in the work of Fama and French (1992, 1995, 1996, and 2008), and Chan, Hamao, and Lakonishok (1991). The sales-to-price support is found in Guerard and Takano (1991) and Ziemba (1992). We refer the reader to those studies and the work of Savita Subramanian at Bank of America Merrill Lynch for testing these variables, and many other strategies in the US equity market. The quantitative work of Subramanian is some of the best "sell side" research, in the opinion of the author.[4]

[4] Savita Subramanian (2011), "US Quantitative Primer," Bank of America Merrill Lynch, May.

In this section, we review and revisit the GPRD regression analysis.[5] Guerard and Takano used relative earnings, book value, cash flow, and sales, relative to sixty month averages, in their analysis. The major papers on combination of value ratios to predict stock returns that include at least CP and/or SP include Chan et al. (1991), Bloch et al. (1993), Lakonishok et al. (1994), and Haugen and Baker (2010). In fact, the Bloch et al. (1993) was a more technical version of Guerard and Takano (1991).

Let us discuss two enhancements in the Guerard et al. (2012) study: the addition of price momentum and earnings per share (eps) forecasts, revisions, and breadth variables. Earnings forecasting enhances returns relative to using only reported financial data and valuation ratios. In 1975, a database of eps forecasts was created by Lynch, Jones, and Ryan, a New York brokerage firm, by collecting and publishing the consensus statistics of 1-year-ahead and 2-year-ahead eps forecasts [Brown (1999)]. The database evolved to become known as the Institutional Brokerage Estimation Service (I/B/E/S) database. There is an extensive literature regarding the effectiveness of analysts' earnings forecasts, earnings revisions, earnings forecast variability, and breadth of earnings forecast revisions, summarized in Bruce and Epstein (1994), Brown (1999), and Ramnath et al. (2008). The vast majority of the earnings forecasting literature in the Bruce and Brown references find that the use of earnings forecasts does not increase stockholder wealth, as specifically tested in Elton et al. (1981) in their consensus forecasted growth variable, FGR. Reported earnings follow a random walk with drift process, and analysts are rarely more accurate than a no-change model in forecasting eps [Cragg and Malkiel (1968)]. Analysts become more accurate as time passes during the year, and quarterly data are reported. Analyst revisions are statistically correlated with stockholder returns during the year [Hawkins et al. (1984) and Arnott (1985)]. Wheeler (1994) developed and tested a strategy in which analyst forecast revision breadth, defined as the number of upward forecast revisions subtracted by the number of downward forecast revisions, divided by the total number of estimates, was the criteria for stock selection. Wheeler found statistically significant excess returns from the breadth strategy. A composite earnings variable, CTEF, is calculated using equally weighted revisions, RV; forecasted earnings yields, FEP; and breadth, BR, of FY1 and FY2 forecasts, a variable put forth in Guerard (1997) and further tested in Guerard, Gultekin, and Stone (1997). Adding I/B/E/S variables in the form of CTEF added to the eight value ratios in

[5] There are many approaches to security valuation and the creation of expected returns. The first approaches to security analysis and stock selection involved the use of valuation techniques using reported earnings and other financial data. Graham and Dodd (1934) recommended that stocks be purchased on the basis of the price-earnings (P/E) ratio and Basu (1977) reported evidence supporting the low P/E model. James (Jim) Miller, Chief Investment Officer, CIO, of Continental Bank commissioned the project with Drexel, Burnham, Lambert, in 1989. Miller and Guerard (1991) presented a stock selection model at The Berkeley Program in Finance that used earnings, book value, cash flow, sales, relative variables, and earnings per share forecast revisions. Miller and Guerard experimented with a price momentum variable, the Columbine Alpha, described in Brush (2001). Jack Brush's Columbine Alpha "pushed out" the eight-factor EP, BP, CP, SP, and relative variables' Efficient Frontier. Guerard delivered paper sat Columbine Equity Research conferences in 1989 and 1994. See Guerard (1990).

Guerard and Takano (1991) produced more than 2.5% of additional annualized return [Guerard et al. (1997)]. The finding of significant predictive performance value for I/B/E/S variables indicates that analyst forecast information has value beyond purely statistical extrapolation of past value and growth measures. Guerard (2012) reported the growing importance of earnings forecasts, revisions, and breadth in Global and the USA, particularly with respect to smaller capitalized securities. Recently, Ramnath, Rock, and Shane (2008) surveyed the financial forecasting literature. Earnings forecasting is still statistically significant in asset selection.

Momentum investing was studied by academics at about the same time that earnings forecasting studies were being published. Levy (1967), Arnott (1979), and Brush and Boles (1983) found statistically significant power in relative strength. The Brush and Boles analysis was particularly valuable because it found that the short-term (3-month) financial predictability of a naïve monthly price momentum model, taking the price at time $t - 1$ divided by the price 12 months ago, $t - 12$, was as statistically significant at identifying underpriced securities as using the alpha of the market model adjusted for the security beta. Brush and Boles found that beta adjustments slightly enhanced the predictive power in the 6–12-month periods. Brush (2001) is an excellent 20-year summary of the price momentum literature. Fama and French (1992, 1995) used a price momentum variable using the price 2 months ago divided by the price 12 months ago, thus avoiding the well-known return or residual reversal effect. The Brush et al. (2004) and Fama studies find significant stock price anomalies, even with Korajcyk and Sadka using transactions costs. The vast majority find that the use of 3-, 6-, and 12-month price momentum variables, often defined as intermediate-term variables, is statistically significantly associated with excess returns.

Guerard et al. (2012) added a Brush-based price momentum: taking the price at time $t - 1$ divided by the price 12 months ago, $t - 12$, denoted PM, and the consensus analysts' earnings forecasts and analysts' revisions composite variable, CTEF, to the stock selection model, one can estimate an expanded stock selection model to use as an input to an optimization analysis. The stock selection model estimated in this chapter, denoted as the United States Expected Returns, USER, is

$$\begin{aligned} TR_{t+1} = a_0 + a_1 EP_t + a_2 BP_t + a_3 CP_t + a_4 SP_t + a_5 REP_t + a_6 RBP_t \\ + a_7 RCP_t + a_8 RSP_t + a_9 CTEF_t + a_{10} PM_t + e_t, \end{aligned} \tag{4.13}$$

where:

EP = [earnings per share]/[price per share] = earnings–price ratio;
BP = [book value per share]/[price per share] = book–price ratio;
CP = [cash flow per share]/[price per share] = cash flow–price ratio;
SP = [net sales per share]/[price per share] = sales–price ratio;
REP = [current EP ratio]/[average EP ratio over the past 5 years];
RBP = [current BP ratio]/[average BP ratio over the past 5 years];
RCP = [current CP ratio]/[average CP ratio over the past 5 years];
RSP = [current SP ratio]/[average SP ratio over the past 5 years];
CTEF = consensus earnings-per-share I/B/E/S forecast, revisions, and breadth;

PM = Price Momentum; and
e = randomly distributed error term.

The composite models could be created by combining variables using OLS, outlier-adjusted or robust regression (ROB), or weighted latent root regression (WLRR) modeling, in which outliers and the high correlations among the variables are used in the estimation procedure. The reader is referred to Bloch et al. (1993) for a discussion of ROB and WLRR techniques.[6] The Markowitz group found that the use of the more advanced statistical techniques produced higher relative out-of-sample portfolio geometric returns and Sharpe ratios. Statistical modeling is not just fun, but it is also consistent with maximizing portfolio returns. The quarterly estimated models outperformed the semiannual estimated models, although the underlying data was semiannual in Japan. The dependent variable in the composite model is total security quarterly returns and the independent variables are the EPR, BPR, CPR, and SPR variables. The ultimate test of OLS, ROB, and WLRR analyses can be found in the Bloch, Guerard, Markowitz, Todd, and Xu (1993) simulations which reported higher Geometric Means, Sharpe Ratios, and F-Statistics using WLRR than OLS in estimating models of the determinants of monthly security returns. The Bloch et al. research (1993) has been reestimated, updated, and enhanced in Guerard (2006), Stone and Guerard (2010), and Guerard, Xu, and Gultekin (2012).

The USER model is estimated using WLRR analysis in (4.13) to identify variables statistically significant at the 10% level; uses the normalized coefficients as weights; and averages the variable weights over the past 12 months. The 12-month smoothing is consistent with the four-quarter smoothing in Guerard and Takano (1991) and Bloch, Guerard, Markowitz, Todd, and Xu (1993).

While EP and BP variables are significant in explaining returns, the majority of the forecast performance is attributable to other model variables, namely, the relative earnings-to-price, relative cash-to-price, relative sales-to-price, price momentum variable, and earnings forecast variable. The consensus earnings forecasting variable, CTEF, and the price momentum variable, PM, dominate the composite model, as is suggested by the fact that the variables account for 45% of the model average weights.

Earnings forecasts, revisions, and directions of revisions are key variables in stock selection modeling. The asset selection of the CTEF variable is highly significant, see Guerard (2012). The average four-quarter smoothed regression coefficients are:
Time-average value of estimated coefficients:

a_1	a_2	a_3	a_4	a_5	a_6	a_7	a_8	a_9	a_{10}
0.044	0.038	0.020	0.038	0.089	0.086	0.187	0.122	0.219	0.224

[6] Guerard (2006) reestimated the GPRD model using PACAP data at The Wharton School from Wharton Research Data Services (WRDS). The WRDS/PACAP data is as close to the GPRD data as was possible in academia. The average cross-sectional quarterly WLRR model F-statistic in the GPRD analysis was 16 during the 1974–1990 period whereas the corresponding F-statistic reported in the Guerard (2006) was 11 for the post-publication, 1993–2001 period. Both sets of models were highly statistically significant and could be effectively used as stock selection models.

Table 4.4 OLS NREG0801 the REG procedure model: MODEL1-dependent variable: RET0801

Number of observations read	3,656
Number of observations used	3,482
Number of observations with missing values	174

Analysis of variance

Source	DF	Sum of squares	Mean square	*F* value	Pr > *F*
Model	10	256.20661	25.62066	28.53	<0.0001
Error	3,471	3,117.52880	0.89816		
Corrected total	3,481	3,373.73542			
Root MSE	0.94772	*R*-square	0.0759		
Dependent mean	0.01606	Adj *R*-sq	0.0733		
Coeff Var	5,899.57118				

Parameter estimates

Variable	DF	Parameter estimate	Standard error	*t* value	Pr > |*t*|
Intercept	1	0.01391	0.01606	0.87	0.3867
EP0801	1	0.18965	0.06321	3.00	0.0027
BP0801	1	−0.01773	0.03834	−0.46	0.6437
CP0801	1	−0.15718	0.07192	−2.19	0.0289
SP0801	1	0.01553	0.04074	0.38	0.7031
REP0801	1	0.01093	0.01573	0.69	0.4873
RBP0801	1	0.01767	0.01807	0.98	0.3283
RCP0801	1	0.02961	0.01579	1.87	0.0609
RSP0801	1	0.14622	0.02064	7.08	<0.0001
CTEF0801	1	0.11279	0.02995	3.77	0.0002
PM0801	1	−0.16049	0.02055	−7.81	<0.0001

In terms of information coefficients, ICs, the use of the WLRR procedure produces the higher IC for the models during the 1998–2007 time period, 0.043, versus the equally weighted IC of 0.040, a result consistent with the previously noted studies.

Let us examine the WLRR SAS output for estimating (4.13) using OLS, ROB using the Beaton–Tukey approximation, and the WLRR techniques for the month of January 2008.

The EP, RCP, RSP, and CTEF variables have the (correct) positive coefficients and are statistically significant in the OLS regression, having t-values that exceed 1.645, the critical 10% level; see Table 4.4. The regression *F*-statistic of 28.53 indicates that the overall regression is highly statistically significant for the 3,482 firm sample in January 2008. The adjusted *R*-squared statistic of 0.073 is quite high for cross-sectional regressions (across securities, at one point in time). The *F*-Statistic of 28.53 is statistically significant at the 1% level. The estimated OLS regression is plagued by outliers, as one sees in Fig. 4.1. The studentized residuals, RStudent, discussed in Chapter 2 and shown in Fig. 4.1, indicate the presence of outliers. A scaled residual known as the Cook distance measure, CookD, or Cook's D, also is shown in Fig. 4.1 and confirms the RStudent result.

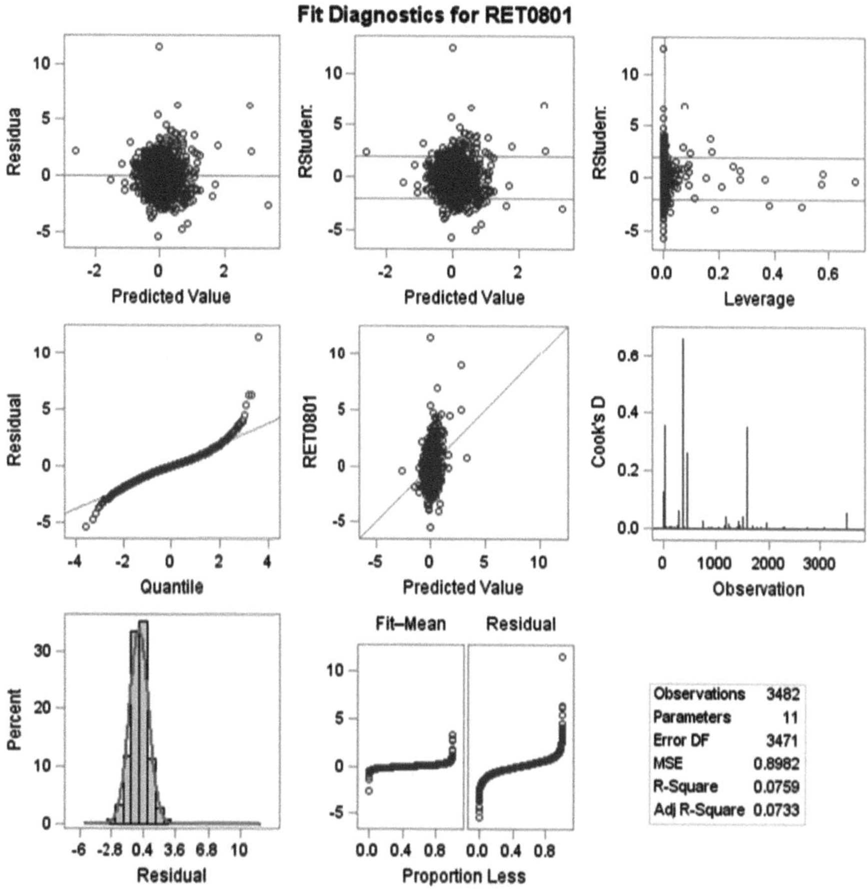

Fig. 4.1 OLS regression diagnostics

Most of the USER variables are associated with OLS outliers, see Fig. 4.2. The BP, CP, SP, RSP, and PM variables are particularly associated with outliers in the January 2008 regression, Fig. 4.3.

The application of the Beaton–Tukey (BT) outlier-adjustment procedure, used in Bloch et al. (1993), increases the F-Statistics from its OLS value of 28.53 to 34.22. Please see Table 4.5. The BT procedure produces positive and statistically significant coefficients on the EP, RSP, and EF (CTEF) variables. The BT procedure reduces the studentized residuals and Cook's D calculated values. Thus, the effect of outliers has been substantially reduced by the Beaton–Tukey Robust Regression application.

The application of the principal components regression analysis, WIPC, in the SAS proc IML procedure approximates of Bloch et al. WLRR. The WIPC

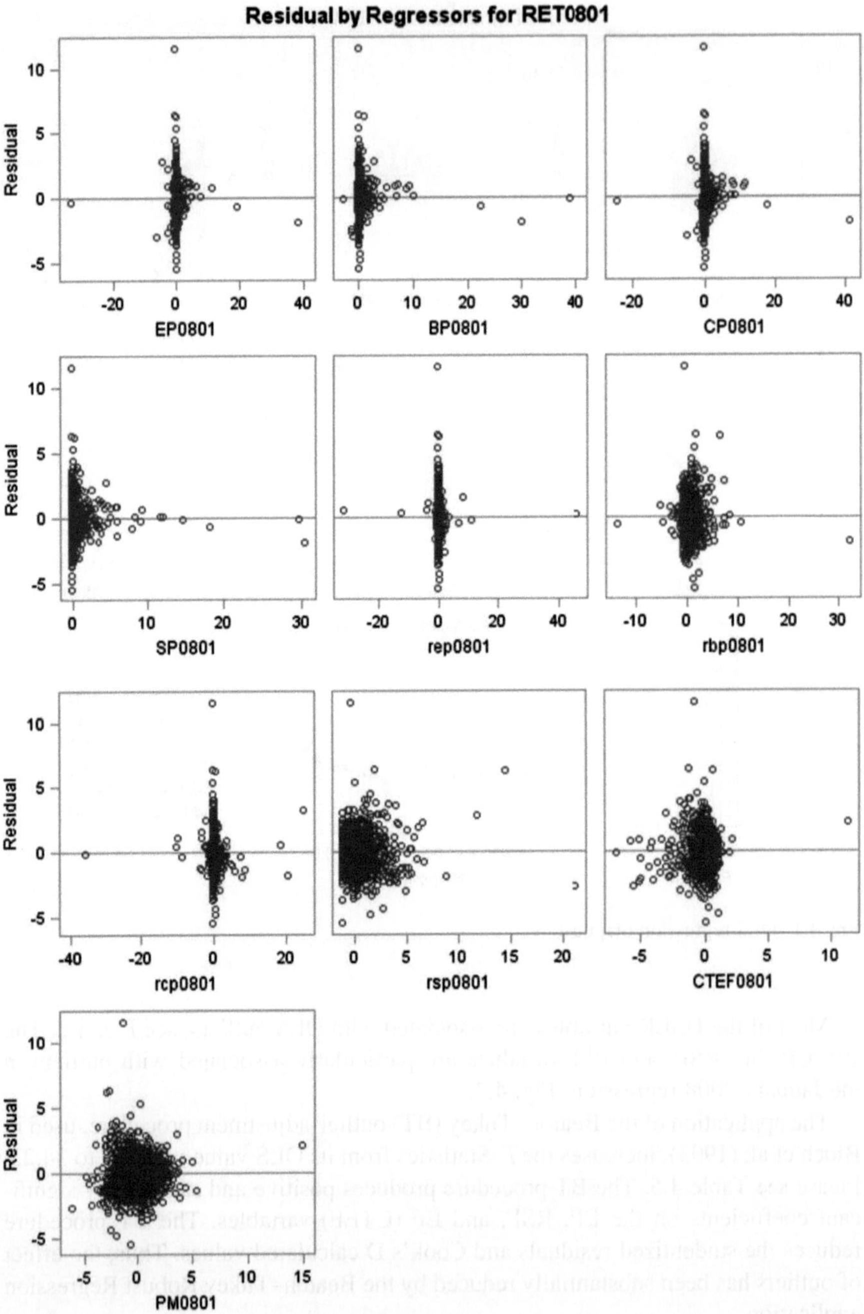

Fig. 4.2 OLS residuals by independent variables

regression analysis shows that the weighted EP, CP, RSP, and CTEF variables are highly statistically significantly associated with security returns in January 2008.
WRDS WIPC 0801

VARN	PC9S	TPC9
WEP0801	0.044	4.618
WBP0801	−0.023	−3.112
WCP0801	0.035	4.506
WSP0801	−0.020	−2.672
WREP0801	0.011	0.992
WRBP0801	0.008	0.489
WRCP0801	0.018	1.352
WRSP0801	0.127	6.615
WEF0801	0.138	5.462
WPM0801	−0.190	−9.768

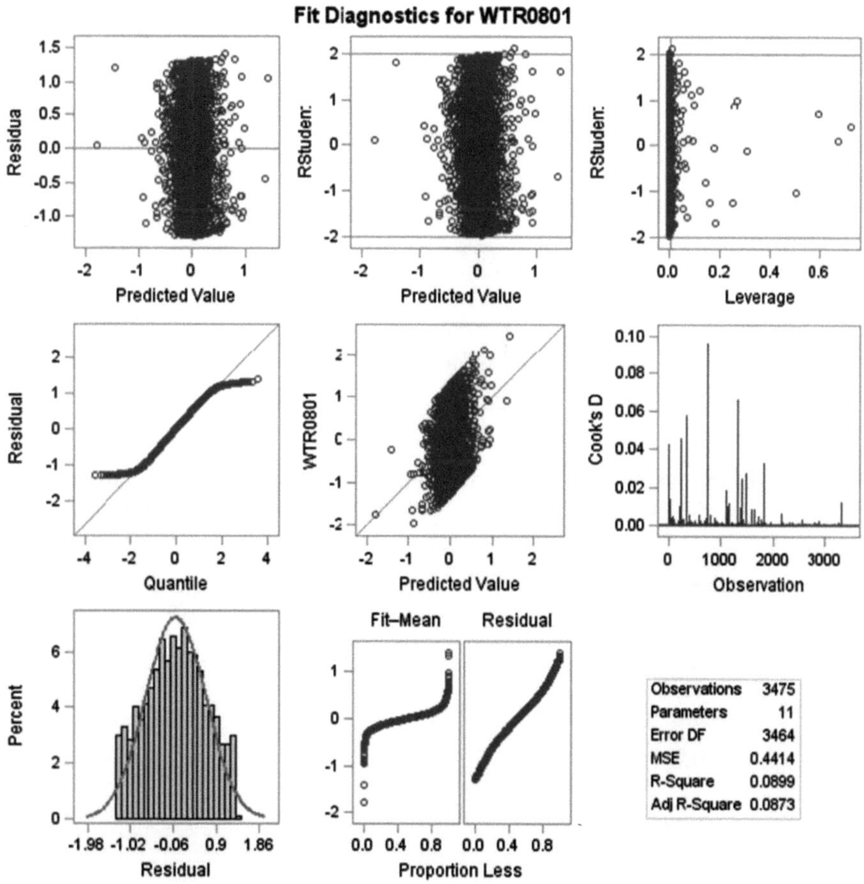

Fig. 4.3 Robust regression diagnostics-dependent variable: WTR0801

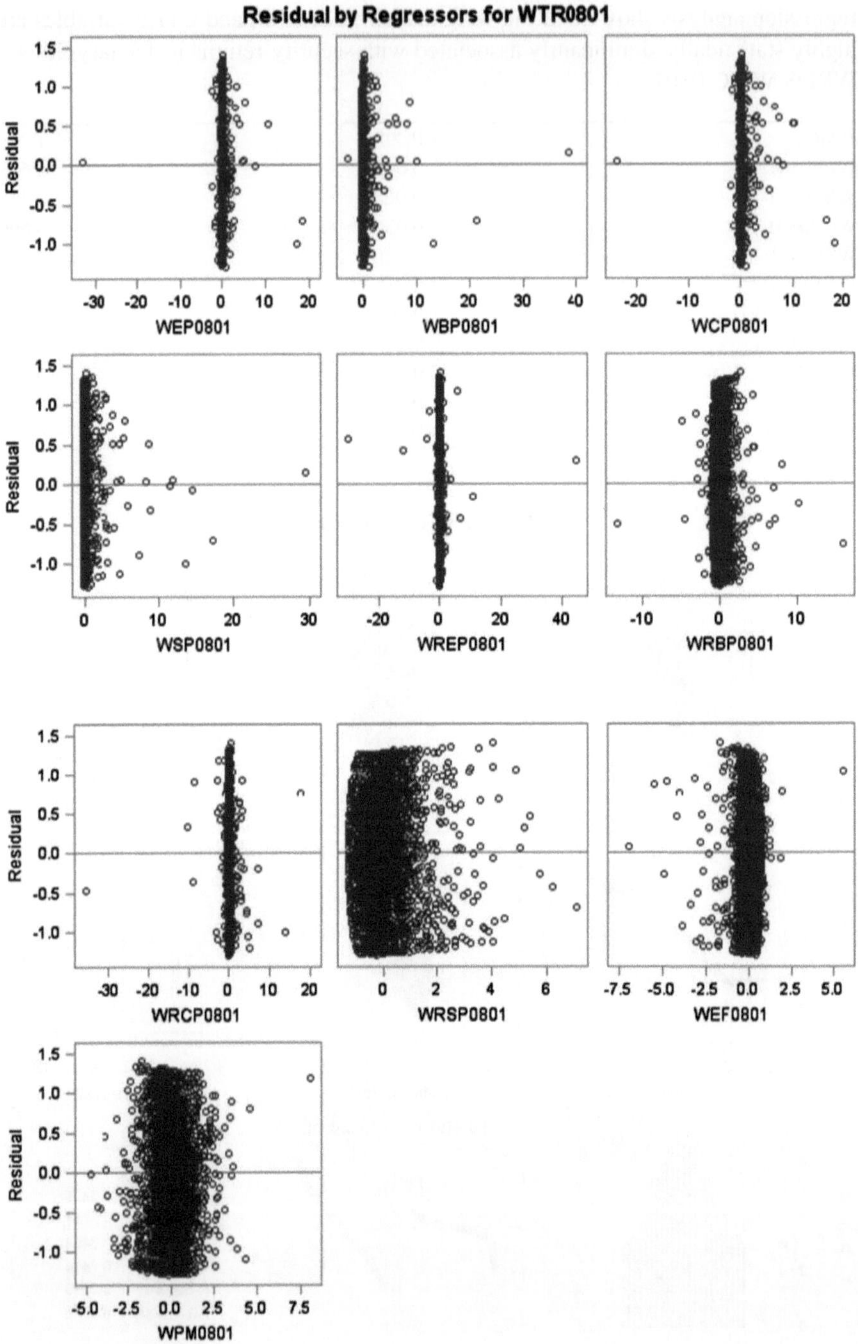

Fig. 4.3 (continued)

The F-Statistic of ROB exceeds the OLS F-Statistic approximately 90% of the months. The ultimate test of OLS, ROB, and WLRR analyses can be found in the Bloch et al. simulations which report higher Geometric Means, Sharpe Ratios, and F-Statistics using WLRR than OLS in estimating models of the determinants of monthly security returns. Moreover, regression weighting of variables outperformed equally weighting the variable in security returns models. We have briefly surveyed the academic literature on anomalies and found substantial evidence that valuation, earnings expectations, and price momentum variables are significantly associated with security returns. Further evidence on the anomalies is found in Levy (1999).[7] We will create portfolios with the USER Model in Chapter 6 and explore more regression modeling of global returns in Chapter 7.

Summary and Conclusions

We have used two case studies to illustrate the effectiveness of regression modeling. Regression analysis offered marginal improvement in the case of combining GNP forecasts, but offered substantial improvement in identifying financial variables associated with security returns. Regression models addressing outliers and multicollinearity problems outperformed equally weighted strategies in stock selection modeling.

---

[7] Haugen and Baker (2010) extended their 1996 study in a recent volume to honor Harry Markowitz. Haugen and Baker estimated their model using weighted least squares. In a given month they estimated the payoffs to a variety of firm and stock characteristics using a weighted least squares multiple regression in each month in the period 1963 through 2007. Haugen and Baker found the most significant factors were; residual Return is last month's residual stock return unexplained by the market.

- Cash Flow-to-Price is the 12-month trailing cash flow-per-share divided by the current price.
- Earnings-to-Price is the 12-month trailing earnings-per-share divided by the current price.
- Return on Assets is the 12-month trailing total income divided by the most recently reported total assets.
- Residual Risk is the trailing variance of residual stock return unexplained by market return.
- 12-month Return is the total return for the stock over the trailing 12 months.
- Return on Equity is the 12-month trailing eps divided by the most recently reported book equity.
- Volatility is the 24-month trailing volatility of total stock return.
- Book-to-Price is the most recently reported book value of equity divided by the current market price.
- Profit Margin is 12-month trailing earnings before interest divided by 12-month trailing sales.
- 3-month Return is the total return for the stock over the trailing 3 months.
- Sales-to-Price is 12-month trailing sales-per-share divided by the market price.

The four measures of cheapness are found in the USER model: cash-to-price, earnings-to-price, book-to-price, and sales-to-price, all have significant positive payoffs. Haugen and Baker (2010) found statistically significant results for the four fundamental factors as did the previously studies we reviewed. The Haugen and Baker (2010) analysis and results are consistent with the Bloch et al. (1993) model.

Table 4.5 ROB NREG0801 the REG procedure model: MODEL1-dependent variable: WTR0801

Number of observations read		3,475			
Number of observations used		3,475			
Analysis of variance					
Source	DF	Sum of squares	Mean square	*F* value	Pr > *F*
Model	10	151.05712	15.10571	34.22	<0.0001
Error	3,464	1,529.11683	0.44143		
Corrected total	3,474	1,680.17395			
Root MSE	0.66440	*R*-square	0.0899		
Dependent mean	0.00118	Adj *R*-sq	0.0873		
Coeff Var	56,310				
Parameter estimates					
Variable	DF	Parameter estimate	Standard error	*t* value	Pr > |*t*|
Intercept	1	0.00627	0.01128	0.56	0.5781
WEP0801	1	0.15940	0.04797	3.32	0.0009
WBP0801	1	−0.02611	0.02782	−0.94	0.3480
WCP0801	1	−0.10215	0.05619	−1.82	0.0691
WSP0801	1	0.01029	0.03011	0.34	0.7327
WREP0801	1	0.01144	0.01141	1.00	0.3164
WRBP0801	1	0.00664	0.01640	0.41	0.6855
WRCP0801	1	0.01746	0.01305	1.34	0.1811
WRSP0801	1	0.12725	0.01919	6.63	<0.0001
WEF0801	1	0.13858	0.02527	5.48	<0.0001
WPM0801	1	−0.16475	0.01690	−9.75	<0.0001

References

Arnott, R. 1979. “Relative Strength Revisited”. *Journal of Portfolio Management*, 19–23.

Arnott, R.1985. “The Use and Misuse of Consensus Earnings.” Journal of Portfolio Management, 18–27.

Armstrong, J.S., *Long Range Forecasting: From Crystal Ball to Computer*, 2nd edition New York: Wiley, 1985.

Ashton, A.H. and Ashton, R.H. 1985. “Aggregating Subjective Forecasts: Some Empirical Results”. *Management Science* 31, 1499–1508.

Basu, S. 1977. “Investment Performance of Common Stocks in Relations to their Price Earnings Ratios: A Test of Market Efficiency.” *Journal of Finance* 32, 663–682.

Bates, J.M. and Granger, C.W.J.1969. “The Combination of Forecasts”. *Operations Research Quarterly* 20 , 451–468.

Beaton, A.E. and J. W. Tukey. 1974. “The Fitting of Power Series, Meaning Polynomials, Illustrated on Bank-Spectroscopic Data,” *Technometrics* 16, 147–185.

Belsey, D.A., 1984. “Collinearity and Forecasting”. *Journal of Forecasting*, 3, 183–96.

Belsey, D.A., Kuh, E. and Welsch, R.E. 1980. *Regression Diagnostics*, New York: Wiley.

Belsley, D.A. 1982. “Assessing the Presence of Harrmful Collinearity and other Forms of Weak Data through a Test for Signal-to-Noise”. *Journal of Econometrics* 20, 211–53.

Blin, J.M. S. Bender, and J. B.Guerard, Jr. 1995. “Why Your Next Japanese Portfolio Should be Market-Neutral- And How”. Journal of Investing 4, 33–40.

Blin, J.M. S. Bender, and J. B.Guerard, Jr. 1997. “Earnings Forecasts, Revisions and Momentum in the Estimation of Efficient Market-Neutral Japanese and U.S. Portfolios.” In *Research in Finance* 15, edited by A. Chen.

Bloch, M., J.B. Guerard, Jr, H. M. Markowitz, P. Todd, and G-L. Xu. 1993. "A Comparison of Some Aspects of the U.S. and Japanese Equity Markets." *Japan & the World Economy* 5, 3–26.

Bloch, M., J.B. Guerard, Jr, H. M. Markowitz, P. Todd, and G-L. Xu. 1993. "A Comparison of Some Aspects of the U.S. and Japanese Equity Markets." *Japan &. the World Economy* 5, 3–26.

Brown, L.D. 1999 and 2008. Annotated I/B/E/S Bibliography.

Brown, L.D. 1999 and 2008. Annotated I7B/E/S Bibliography.

Bruce, B. and C.B. Epstein. 1995. *The Handbook of Corporate Earnings Analysis*. Chicago: Probus.

Brush, J. S. 2007. "A Flexible Theory of Price Momentum". *Journal of Investing* 16.

Brush, J. S. 2007. "A Flexible Theory of Price Momentum". *Journal of Investing* 16.

Brush, J. S. and K.E. Boles. 1983. "The Predictive Power in Relative Strength and CAPM". *Journal of Portfolio Management,* 20–23.

Brush, J. S. and K.E. Boles. 1983. "The Predictive Power in Relative Strength and CAPM". *Journal of Portfolio Management*, 20–23.

Brush, J.S. 2001."Price Momentum: A Twenty-Year Research Effort". Columbine Newsletter Special Issue.

Brush, J.S. 2001."Price Momentum: A Twenty-Year Research Effort". Columbine Newsletter Special Issue.

Bunn, D. W. 1989. "Forecasting with more than One Model". *Journal of Forecasting* 8, 161–166.

Bunn, D.W. 1975. "A Bayesian Approach to the Linear Combination of Forecasts", *Operational Research Quarterly* 26, 325–329.

Chan, L.K.C., J. Lakonishok, and Y. Hamao 1993. "Can Fundamentals Predict Japanese Stock Returns" *Financial Analysts Journal.*

Chan, L.K.C., N. Jegadeesh and J. Lakonishok. 1996. "Momentum Strategies." *Journal of Finance* 51, 1681–1713.

Chan, L.K.C., Y. Hamao, and J. Lakonishok. 1991. "Fundamentals and Stock Returns in Japan". *Journal of Finance* 46, 1739–1764.

Chan, L.K.C., Y. Hamao, and J. Lakonishok. 1991. "Fundamentals and Stock Returns in Japan". *Journal of Finance* 46, 1739–1764.

Chen, N-F., R. Roll, and S. Ross. 1986. "Economic Forces and the Stock Market." *Journal of Business* 59, 383–403.

Clemen, R. T. and J.B. Guerard, Jr. 1989. "Econometric GNP Forecasts: Incremental Information Relative to Naïve Extraploation." *International Journal of Forecasting* 5, 417–426.

Clemen, R.T. 1989. "Combining Forecasts: A Review and Annotated Bibliography". *International Journal of Forecasting* 5, 559–583.

Clemen, R.T. and Winkler, R.L. 1986, "Combining Economic Forecasts". *Journal of Business and Economic Statistics* 4, 39–46.

Clemen, R.T., 1986. "Linear Constraints and the Efficiency of Combined Forecasts". *Journal of Forecasting* 5, 31–38.

Cragg, J.G. and B.G. Malkiel.1968. "The Consensus and Accuracy of Some Predictions of the Growth of Corporate Earnings". *Journal of Finance* 23, 67–84.

Dimson, E. 1988. *Stock Market Anomalies.* Cambridge: Cambridge University Press.

Doornik, J. A. and D. F. Hendry. 2009. Empirical Econometric Modelling. Timberlake Consultants, Ltd.

Doornik, J. A. and D. F. Hendry. 2009. Modelling Dynamic Systems. Timberlake Consultants, Ltd.

Dremen, D. 1998. *Contrarian Investment Strategies: The Next Generation.* New York: Simon &. Schuster.

Elton, E. J. and M.J. Gruber. 1972. "Earnings Estimates and the Accuracy of Expectational Data." *Management Science* 18, B409–B424.

Elton, E. J., M.J. Gruber, and M. Gultekin. 1981. "Expectations and Share Prices." *Management Science* 27, 975–987.

Elton, E. J., M.J. Gruber, and M. Gultekin. 1981. "Expectations and Share Prices." *Management Science* 27, 975–987.

Elton, E.J., M.J. Gruber, S.J. Brown, and W.N. Goetzman. 2007. *Modern Portfolio Theory and Investment Analysis.* John Wiley &. Sons, Inc., Seventh Edition.

Fama, E.F. and J.D. MacBeth. 1973. "Risk, Return, and Equilibrium: Empirical Tests." *Journal of Political Economy* 81, 607–636.
Fama, E.F. and K. R. French. 1992. "Cross-Sectional Variation in Expected Stock Returns." *Journal of Finance* 47, 427–465.
Fama, E.F. and K. R. French. 2008. "Dissecting Anomalies". *Journal of Finance* 63, 1653–1678.
Fama, E.F. and K.R. French. 1995. "Size and the Book-to-Market Factors in Earnings and Returns. *Journal of Finance* 50, 131–155.
Fama, E.F., and K.R. French. 1996. "Multifactor Explanations of Asset Pricing Anomalies". *Journal of Finance* 51, 55–84.
Farrell, J.L. Jr. 1974. "Analyzing Covariance of Returns to Determine Homogeneous Stock Groupings." *Journal of Business* 47, 186–207.
Farrell, J.L. Jr. 1997. *Portfolio Management: Theory and Applications.* New York McGraw-Hill/Irwin.
Ferson, W.E. and C.R. Harvey. 1998. "Fundamental Determinants of National Equity Market Returns: A Perspective on Conditional Asset Pricing." *Journal of Banking &. Finance* 21, 1625–1665.
Figlewski, S. and Urich, T. 1983. "Optimal Aggregation of Money Supply Forecasts: Accuracy, Profitability and Market Efficiency". *Journal of Finance* 28, 695–710.
Fisher, L. 1966a. "An Algorithm for Finding Exact Rates of Return." *Journal of Business* 39, 111–118.
Fisher, L. 1966b. "Some New Stock-Market Indexes." *Journal of Business* 39, 191–255.
Fisher, L. and J. H. Lorie. 1964. "Rates of Return on Investments in Common Stocks." *Journal of Business* 7, 1–21.
Fisher, L. and J. H. Lorie. 1970. "Some Studies of Variability of Returns on Investments in Common Stocks." *Journal of Business* 43, 39–143.
Graham, B., D. Dodd, and S. Cottle. 1934, 1962. *Security Analysis: Principles and Technique.* New York: McGraw-Hill Book Company. Editions one and four.
Granger, C.W.J. 1989. "Invited Review: Combining Forecasts-Twenty Years Later". *Journal of Forecasting* 8, 167–173.
Granger, C.W.J. and Newbold, P. 1977. *Forecasting Economic Time Series,* New York: Academic Press.
Granger, C.W.J. and Ramanathan, R., 1984. "Improved Methods of Combining Forecasts", *Journal of Forecasting*, 3 (1984), 197–204.
Grinold, R and R. Kahn. 1999. *Active Portfolio Management New* York: McGraw-Hill/Irwin.
Guerard, J.B. , Jr. 2012. "Global Earnings Forecast Efficiency" in J. Kensinger, Ed., Research in Finance 28, 19–47.
Guerard, J.B. 1997. "Is There a Cost to be Socially Responsible in Investing?" *Journal of Forecasting* 16, 475–490.
Guerard, J.B. 2012. Global Earnings Forecast Efficiency (2012), *Research in Finance* 28, 19–47.
Guerard, J.B., Jr. 1990. "Optimal Portfolio Analysis using Composite Model-Tilting". Drexel Burnham, Lambert. Unpublished working paper.
Guerard, J.B., Jr. 2006. "Quantitative Stock Selection in Japan and the US: Some Past and Current Issues." *Journal of Investing,* 15.
Guerard, J.B., Jr. and M. Takano. 1991. "The Development of Mean-Variance Portfolios in Japan and the U.S.". *Journal of Investing* 1.
Guerard, J.B., Jr. and R.T. Clemen. 1989. "Collinearity and the Use of Latent Root Regression for Combining GNP Forecasts". *Journal of Forecasting* 8, 231–238.
Guerard, J.B., Jr., M. Gultekin, and B.K. Stone. 1997. "The Role of Fundamental Data and Analysts' Earnings Breadth, Forecasts, and Revisions in the Creation of Efficient Portfolios." In *Research in Finance* 15, edited by A. Chen.
Guerard, J.B., Jr., M.N. Gultekin, and G. Xu. 2012. "Investing with Momentum: The Past, Present, and Future", *Journal of Investing* 21, 68–80.
Guerard, J.B., Jr., M.N. Gultekin, and G. Xu. 2012. "Investing with Momentum: The Past, Present, and Future", *Journal of Investing* 21, 68–80.

Guerard, J.B., Jr., S. C. Chettiappan, and G-L, Xu. 2010. "Stock-Selection Modeling and Data Mining Corrections: Long-Only versus 130/30 Models". *The Handbook of Portfolio Construction: Contemporary Applications of Markowitz Techniques*. New York: Springer.

Gunst, R.F., J.T. Webster, and R.L. Mason. 1976. "A Comparison of Least Squares and Latent Root Regression Estimators." *Technometrics* 18, 75–83.

Gunst, R.F., Webster, J.T. and Mason, R.L. 1976. "A Comparison of Least Squares and Latent Root Regression Estimators". *Technometrics*, 18, 75–83.

Gupta, S. and Wilton, P.C., "Combination of Forecasts: An Extension", *Management Science*, 33 (1987), 356–72.

Haugen, R. and N. Baker. 2010. "Case Closed". *The Handbook of Portfolio Construction: Contemporary Applications of Markowitz Techniques*. New York: Springer.

Haugen, R. and N. Baker. 2010. "Case Closed". *The Handbook of Portfolio Construction: Contemporary Applications of Markowitz Techniques*. New York: Springer.

Hawawini, G. and D.B. Keim. 1995. "On the Predictability of Common Stock Returns: World-Wide Evidence", in R.A. Jarrow, V. Maksimovic, and W.T. Ziemba, Editors, *Handbooks in Operations Research and Management Science: Finance* 9, 497–537.

Hocking, R.R. 1976. "The Analysis and Selection of Variables in Linear Regression". *Biometrics* 32 , 1–49.

Kang, H. 1986. "Unstable Weights in the Combination of forecasts". *Management Science* 32, 683–95.

Keane, M.P. and D.E. Runkle. 1998. "Are Financial Analysts' Forecasts of Corporate Profits Rational?" *The Journal of Political Economy* 106, 768–805.

Korajczyk, R.A. and R. Sadka. 2004. "Are Momentum Profits Robust to Trading Costs?" *Journal of Finance* 59 (2004), 1039–1082.

Lakonishok, J., A. Shleifer, and R.W. Vishny. 1994. "Contrarian Investment, Extrapolation and Risk." *Journal of Finance* 49, 1541–1578.

Latane, H. A., D. L. Tuttle, and CP. Jones. 1975. *Security Analysis and Portfolio Management*. New York: The Roland Press. 2nd Edition.

Latane, H.A. 1959."Criteria for Choice Among Risky Ventures". *Journal of Political Economy* 67, 144–155.

Lesmond, D.A., M.J. Schill, and C. Zhou. 2004. "The Elusory Nature of Trading Profits". *Journal of Financial Economics* 71, 349–380.

Levy, H. 1999. Introduction to Investments. Cincinnati: South-Western College Publishing. Second Edition.

Lim, T. 2001. "Rationality and Analysts' Forecast Bias". *Journal of Finance* 56, 369–385.

Lindley, D.V. and Smith, A.F.M. 1972. "Bayes estimates for the linear model". *Journal of the Royal Statistical Society*, Series B, 34, 1–41.

Lintner, J. 1965a. "The Valuation of Risk Assets on the Selection of Risky Investments in Stock Portfolios and Capital Investments." *The Review of Economics and Statistics*, 13–37.

Lintner, J. 1965b. "Security Prices, Risk, and the Maximum Gain from Diversification." *Journal of Finance* 30.

Makridakis, S. and Winkler, R.L. 1983. "Averages of Forecasts: Some Empirical Results". *Management Science* 29, 987–996.

Markowitz, H. M. 1956. The Optimization of a Quadratic Function Subject to Linear Constraints. *Naval Research Logistics Quarterly* 3, 111–133.

Markowitz, H.M. 1952. "Portfolio Selection." *Journal of Finance* 7, 77–91.

Markowitz, H.M. 1952. "Portfolio Selection." *Journal of Finance* 7, 77–91.

Markowitz, H.M. 1959. *Portfolio Selection: Efficient Diversification of Investment*. Cowles Foundation Monograph No. 16. New York, John Wiley &. Sons.

Markowitz, H.M. 1959. *Portfolio Selection: Efficient Diversification of Investment*. Cowles Foundation Monograph No.16. New York, John Wiley & Sons.

Markowitz, H.M. 1976. "Investment in the Long Run: New Evidence for an Old Rule," *Journal of Finance* 31, 1273–1286.

Markowitz, H.M. 2000. *Mean-Variance Analysis in Portfolio Choice and Capital Markets.* New Hope, PA: Frank J. Fabozzi Associates.
Markowitz, H.M. *Mean-Variance Analysis in Portfolio Choice and Capital Markets.* Oxford: Basil Blackwell, 1987.
Markowitz, H.M. and G.L. Xu. 1994. "Data Mining Corrections". *Journal of Portfolio Management.* Fall.
Markowitz, H.M., and G. Xu. 1994. "Data Mining Corrections." *Journal of Portfolio Management* 21, 60–69.
Miller, J. D., J.B. Guerard, Jr. and M. Takano. 1991. "Bridging the Gap Between Theory and Practice in Equity Selection Modeling: Case Studies of U.S. and Japanese Models. An earlier version was presented at the Berkeley Program in Finance. September 1990. Santa Barbara, CA.
Mincer, J. and V. Zarnowitz. (1969). "The Evaluation of Economic Forecasts," J. Mincer editor, *Economic Forecasts and Expectations,* New York, Columbia University Press.
Montgomery, D.C. and Peck, E.A. 1982. *Introduction to Linear Regression Analysis,* New York: Wiley.
Mossin, J. 1966. "Equilibrium in a Capital Asset Market." *Econometrica* 34, 768–783.
Nerlove, M. 1968. "Factors Affecting Differences Among Rates of Return on Investments in Individual Common Stocks". *Review of Economics and Statistics* 50, 312–331.
Newbold, P. and Granger, C.W.J. 1974. "Experience with Forecasting Univarate Time Series and the Combination of Forecasts". *Journal of the Royal Statistical Society*, Series A., 137, 131–164.
Rachev, S., B. Rachev-Itova, S Stoyanov, and F. Fabozzi. 2010. "Risk Management and Portfolio Optimization for Volatile Markets", in Guerard, J.B., editor, *The Handbook of Portfolio Construction: Contemporary Applications of Markowitz Techniques,* New York: Springer.
Ramnath, S., S. Rock, and P. Shane. 2008. "The Financial Analyst Forecasting Literature: A Taxonomy with Suggestions for Further Research. *International Journal of Forecasting* 24, 34–75.
Rudd, A. and B. Rosenberg. 1980. "The 'Market Model' in Investment Management." *Journal of Finance* 5, 597–607.
Sharma, S. and James, W.L. 1981. "Latent Root Regression: An Alternative Procedure for Estimating Parameters in the Presence of Multicollinearity". *Journal of Market Research* 18, 154–161.
Stone, B. K. and J.B. Guerard, Jr. 2010a. Methodologies for Isolating and Assessing the Portfolio Performance Potential of Stock Return Forecast Models with an Illustration, in J.B. Guerard, Jr., editor, *The Handbook of Portfolio Construction: Contemporary Applications of Markowitz Techniques.* New York: Springer.
Stone, B. K. and J.B. Guerard, Jr. 2010b. "Quantitative Stock Selection in Japan and the United States: Some Past and Current Issues." Proceedings, 50th Annual CRSP Forum (Chicago, November 2010)
Subramanian , S. 2011. "US Quantitative Primer", Bank of America Merrill Lynch, May.
Vander Weide, J. H. 2010. "Principles for Lifetime Portfolio Selection: Lessons from Portfolio Theory", in J. Guerard, Ed. *Handbook of Portfolio Construction: Contemporary Applications of Markowitz Techniques*. New York: Springer.
Webster, J.T., Gunst, R.F. and Mason, R.L. 1974. "Latent Root Regression Analysis". *Technometrics* 16, 513–22.
Wheeler, L.B. 1994. "Changes in Consensus Earnings Estimates and Their Impact on Stock Returns." In *The Handbook of Corporate Earnings Analysis.* Edited by B. Bruce and C. B. Epstein.
Winkler, R.L. and Makridakis, S. 1983. "The Combination of Forecasts". *Journal of the Royal Statistical Society*, Series A., 146, 150–157.
Winkler, R.L., 1981. "Combining Probability Distributions from Dependent Information Sources". *Management Science* 27, 479–88.
Ziemba, W. T. 1990. "Fundamental Factors in US and Japanese Stock Returns." Berkeley Program in Finance.
Ziemba, W. T. 1992. *Invest Japan.* Chicago: Probus Publishing Co.

Chapter 5
Transfer Function Modeling and Granger Causality Testing

In this chapter we fit univariate and bivariate time series models in the tradition of Box and Jenkins (1976) and Granger and Newbold (1977) and apply traditional Granger causality testing following the Ashley et al. (1980) methodology. Second, we estimate Vector Autoregressive Models (VAR) and Chen and Lee (1990) Vector ARMA (VARMA) causality test. We test two series for causality: (1) stock prices and mergers and (2) the money supply and stock prices.

Testing for Causality: The Ashley et al. (1980) Test

There is a large and growing literature on causality testing in economics. Clive Granger, one of the great minds in time series, reminds us that The phrase "*X* causes *Y*" must be handled with considerable delicacy, as the concept of causation is a very subtle and difficult one (Ashley, Granger, and Schmalensee (1980)). We will refer to Ashley, Granger, and Schmatensee (1980) as AGS (1980). Granger held that a universally acceptable definition of causation may well not be possible, but a reasonable definition might be the following: Let Ω_n represent all the information available in the universe at time n. Suppose that at time n optimum forecasts are made of X_{n+1} using all of the information in Ω_n and also using all of this information apart from the past and present values Y_{n-j}, $j \geq 0$, of the series Y_t. If the first forecast, using all the information, is superior to the second, then the series Y_t has some special information about X_t, not available elsewhere, and Y_t is said to cause X_t. Before applying this definition, one must establish the criteria to decide if one forecast is superior to another. The usual procedure is to compare the relative mean-square errors of post-sample forecasts, as we discussed in Chapter 1.

To make the suggested definition suitable for practical use a number of simplifications have to be made. Linear forecasts only will be considered, together with the usual least-squares loss function, and the information set Ω_n has to be

J.B. Guerard, Jr., *Introduction to Financial Forecasting in Investment Analysis*,
DOI 10.1007/978-1-4614-5239-3_5,

replaced by the past and present values of some set of time series, R_n: $\{X_{n-j}, Y_{n-j}, Z_{n-j}, \ldots, j \geq 0\}$. Any causation now found will only be relative to R_n; spurious results can occur if some vital series is not in this set.

The simplest case is when R_n consists of just values from the series X_t and Y_t, where now the definition reduces to the following: let MSE (X) be the population mean-square of the one-step forecast error of X_{m+1} using the optimum linear forecast based on X_{n-j}, $j \geq 0$, and let MSE (X, Y) be the population mean-square of the one-step forecast error of X_{n+1} using the optimum linear forecast based on X_{n-j}, $Y_{n-j} j \geq 0$. Then Y *causes* X if MSE (X, Y) $<$ MSE (X). The testing involving the definition of causation (stated in terms of variances rather than mean-square errors) was introduced into the economic literature by Granger (1969) and it has been applied by Sims (1972) and Ashley et al. (1980), which we will refer to as AGS (1980).

AGS (1980) proposed several step approach to the analysis of causality between a pair of time series X_t and Y_t:

(i) Each series is prewhitened by building single-series ARIMA models using the Box–Jenkins procedure.

(ii) Form the cross-correlogram between these two residual series,

$$\rho_k = \text{corr}(\text{res } x_t, \ \text{res } y_{t-k}).$$

(iii) For positive and negative values of k: If any ρ_k for $k > 0$ are significantly different from zero, there is an indication that Y_t may be causing X_t, since the correlogram indicates that past Y_t may be useful in forecasting X_t. Similarly, if any ρ_k is significantly nonzero for $k < 0$, X_t appears to be causing Y_t. If both occur, two-way causality, or feedback, between the series is indicated. AGS (1980) note that the sampling distribution of the ρ_k depends on the exact relationship between the series. On the null hypothesis of no relationship, it is well known that the ρ_k are asymptotically distributed as independent normal with means zero and variances $1/n$, where n is the number of observations employed, but the experience shows that the test suggested by this result must be used with extreme caution in finite samples.[1] In practice, we also use a priori judgement about the forms of plausible relations between economic time series. Thus for example, a value of ρ_1 well inside the interval $[-2/\sqrt{n}, \ +2/\sqrt{n}]$ might be tentatively treated as significant, while a substantially larger value of ρ_7 might be ignored if ρ_5, ρ_6, ρ_8, and ρ_9 are all negligible.

This step is analogous to the univariate Box–Jenkins identification step, where a tentative specification is obtained by judgmental analysis of a correlogram. The key word is "tentative"; the indicated direction of causation is only tentative at this stage and may be modified or rejected on the basis of subsequent modeling and forecasting results.

[1] One must apparently be even more careful with the Box–Pierce test on sums of squared ρ_k.

(iv) For every indicated causation, a bivariate model relating the residuals is identified, estimated, and diagnostically checked. If only one-way causation is present, the appropriate model is unidirectional and can be identified directly from the shape of the cross-correlogram, see Granger and Newbold (1977).
(v) From the fitted model for residuals, after dropping insignificant terms, the corresponding model for the original series is derived, by combining the univariate models with the bivariate model for the residuals. It is then checked for common factors, estimated, and diagnostic checks applied.[2]
(vi) Finally, the bivariate model for the original series is used to generate a set of one-step forecasts for a post-sample period. The corresponding errors are then compared to the post-sample one-step forecast errors produced by the univariate model developed in step (i) to see if the bivariate model actually does forecast better.[3] The use of sequential one-step forecasts follows directly from the definition above and avoids the problem of error buildup that would otherwise occur as the forecast horizon is lengthened.

Because of specification and sampling error (and perhaps some structural change) the two forecast error series thus produced are likely to be cross-correlated and autocorrelated and to have nonzero means. In light of these problems, no direct test for the significance of improvements in mean-squared forecasting error appears to be available. Consequently, we have developed the following indirect procedure.

For some out-of-sample observation, t, let e_{1_r} and e_{2_r} be the forecast errors made by the univariate and bivariate models, respectively, of some time series. Elementary algebra then yields the following relation among sample statistics for the entire out-of-sample period:

$$\mathrm{MSE}(e_1) - \mathrm{MSE}(e_2) = [s^2(e_1) - s^2(e_2)] + [m(e_1)^2 - m(e_1)^2], \tag{5.1}$$

where MSE denotes sample mean-squared error, s^2 denotes sample variance, and m denotes sample mean. Letting

$$\Delta_t = e_{1t} - e_{2t} \quad \text{and} \quad \sum_2 = e_{1t} + e_{2t}, \tag{5.2}$$

---

[2] OLS estimation suffices to produce unbiased estimates, since all the bivariate models considered are reduced forms. It also allows one to consider variants of one equation without disturbing the forecasting results from the other, and it is computationally simpler. On the other hand, where substantial contemporaneous correlation occurs between the residuals, seemingly unrelated regression GLS estimation can be expected to yield noticeably better parameter estimates and post-sample forecasts. All estimation in this study is OLS; a re-estimation of our final bivariate model using GLS might strengthen our conclusions somewhat.

[3] Alternatively, one might fit both models to the sample period, produce forecasts of the first post-sample observation, reestimate both models with that observation added to the sample, forecast the second post-sample observation, and so on until the end of the post-sample period. This would, of course, be more expensive than the approach in the text.

equation (5.1) can be rewritten as follows, even if e_{1t} and e_{2t} are correlated:

$$\mathrm{MSE}(e_1) - \mathrm{MSE}(e_2) = \left[\widehat{\mathrm{cov}}\left(\Delta, \sum\right)\right] + \left[m(e_1)^2 - m(e_2)^2\right], \tag{5.3}$$

where $\widehat{\mathrm{cov}}$ denotes the sample covariance over the out-of-sample period.

Let us assume that both error means are positive; the modifications necessary in the other cases should become clear. Consider the analogue of (5.3) relating population parameters instead of sample statistics, and let cov denote the population covariance and μ denote the population mean. From (5.3), it is then clear that we can conclude that the bivariate model outperforms the univariate model if we can reject the joint null hypothesis cov $(\Delta, \sum) = 0$ and $\mu(\Delta) = 0$ in favor of the alternative hypothesis that both quantities are nonnegative and at least one is positive.

Now consider the regression equation

$$\Delta_t = \beta_1 + \beta_2\left[\sum_t - m\left(\sum_t\right)\right] + \mu_t, \tag{5.4}$$

where μ_t is an error term with mean zero that can be treated as independent of $\sum_t$. From the algebra of regression, the test outlined in the preceding paragraph is equivalent to testing the null hypothesis $\beta_1 = \beta_2 = 0$ against the alternative that both are nonnegative and at least one is positive. If either of the two least squares estimates, $\hat{\beta}_1$ and $\hat{\beta}_2$, is significantly negative, the bivariate model clearly cannot be judged a significant improvement. If one estimate is negative but not significant, a one-tailed t test on the other estimated coefficient can be used. If both estimates are positive, an F test of the null hypothesis that both population values are zero can be employed. But this test is, in essence, four-tailed; it does not take into account the signs of the estimated coefficients. If the estimates were independent, it is clear that the probability of obtaining an F-statistic greater than or equal to F_0, say, and having both estimates positive is equal to one-fourth the significance level associated with F_0. Consideration of the possible shapes of iso-probability curves for $\left(\hat{\beta}_1, \hat{\beta}_2\right)$ under the null hypothesis that both population values are zero establishes that the true significance level is never more than half the probability obtained from tables of the F distribution. If both estimates are positive, then one can perform an F test and report a significance level equal to half that obtained from the tables.

The approach just described differs from others that have been employed to analyze causality in its stress on models relating the original variables and on post-sample forecasting performance. We now discuss these two differences.

Models directly relating the original variables provide a sounder, as well as a more natural basis for conclusions about causality. As has been argued in detail by Granger and Newbold (1977), however, prewhitening and analysis of the cross-correlogram of the prewhitened series are useful steps in the identification of models relating the original series, since the cross-correlogram of the latter is likely to be impossible to interpret sensibly. Because the correlations between the prewhitened series (the ρ_k) have unknown sampling distributions, this analysis involves subjective judgements, as does the identification step in univariate Box–Jenkins analysis. AGS (1980) state that in neither case is an obviously better approach available, and in both cases the tentative conclusions reached are subjected to further tests.

It is somewhat less clear how out-of-sample data are optimally employed in an analysis of causality. This question is closely related to fundamental problems of model evaluation and validation and is complicated by sampling error and possible specification error and time-varying coefficients. The riskiness of basing conclusions about causality entirely on within-sample performance is reasonably clear. Since the basic definition of causality is a statement about forecasting ability, it follows that tests focusing directly on forecasting are most clearly appropriate. Indeed, it can be argued that goodness-of-fit tests (as opposed to tests of forecasting ability) are contrary in spirit to the basic definition.[4] Moreover, within-sample forecast errors have doubtful statistical properties in the present context when the Box–Jenkins methodology is employed. While the power of that methodology has been demonstrated in numerous applications and rationalizes our use of it here, it must be noted that the identification (model specification) procedures in steps (i)–(iv) above involve consideration and evaluation of a wide variety of model formulation. A good deal of sample information is thus employed in specification choice, and there is a sense in which most of the sample's real degrees of freedom are used up in this process. It thus seems both safer and more natural to place considerable weight on out-of-sample forecasting performance.

The approach outlined above uses the post-sample data only in the final step, as a test track over which the univariate and bivariate models are run in order to compare their forecasting abilities. This approach is of course vulnerable to undetected specification error or structural change. Partly as a consequence of this, the likely characteristics of post-sample forecast errors render testing for performance improvement somewhat delicate, as we noted above. Finally, the appropriate division of the total data set into sample and post-sample periods in the AGS (1980)

[4] If one finds that one model (using a wider information set, say) fits better than another, one is really saying "If I had known that at the beginning of the sample period, I could have used that information to construct better forecasts *during* the sample period." But this is not strictly operational and thus seems somewhat contrary in spirit to the basic definition of causality that we employ.

approach is not a nontrivial problem. Additional work on in-sample and post-sample analysis is reported in Ashley (2003) and Thomakos and Guerard (2004). Our basic point is simply that model specification (perhaps especially within the Box–Jenkins framework) may well be infected by sampling error.

AGS applied their methodology to aggregate advertising and consumption during the 1956–1975 period. The bivariate aggregate consumption model, using aggregate advertising as its input, reduced the out-of-sample forecasting error by only 5.1 % relative to the univariate aggregate consumption model, indicating that aggregate advertising does not cause aggregate consumption. The bivariate aggregate advertising model, using aggregate consumption as its input, reduced the out-of-sample forecasting error by 26 % relative to the univariate aggregate advertising model, indicating that aggregate consumption causes aggregate advertising.

Quarterly Mergers, 1992–2011: Automatic Time Series Modeling and an Application of the Ashley et al. (1980) Test

Let us explore further the AGS (1980) approach using a case study of aggregate mergers using Mergerstat quarterly data from 1992 to 2011. There is a well-established history of mergers and stock prices.[5] Guerard (1985) used the AGS (1980) bivariate transfer function causality testing methodology and reported that stock prices led mergers over the Nelson quarterly data from 1895 to 1954. Guerard reported that the bivariate merger model, with stock prices as its input, reduced the out-of-sample forecasting errors by 35.7 % less than the univariate time series merger model. Thus, quarterly stock prices led mergers over the 1895–1954 period. We use the AGS (1980) approach to model mergers as a function of leading economic indicators (LEI) and stock prices (using the S&P 500). Most economic

[5] The merger history of the United States was studied by Nelson (1959), who reported that mergers were highly correlated with stock prices and industrial production from 1895 to 1954. Nelson (1966) later found that stock prices lead mergers by over 5 months (5.25) over the 1919–1961 period. Melicher et al. (1983) and Guerard (1985) used ARIMA and transfer function modeling to find that stock prices lead mergers. Guerard and McDonald (1995) reported that the annual merger series from 1895 to 1979 was a near-random walk and that outlier-estimated time series models did not statistically outperform the naïve random walk with drift model. Golbe and White (1993) fit a sine wave to a "spliced" US annual merger history and found that a sine wave, representing a 40-year merger model, described the behavior of mergers.

historians recite the major merger movements and their "waves" since 1895.[6] A time series of the US quarterly data is obtained from the FactSet Mergerstat database for 1992–2011Q2. The data is read into Oxmetrics. We run an analysis of the quarterly data in which the change in the logarithmic transformation (dlog) of mergers is a function of the dlog components of the LEI published by The

---

[6] The US merger history was characterized by George Stigler (1950) to have occurred in three waves. The first major merger movement began in 1879, with the creation of the Standard Oil Trust, and ended with the depression of 1904. During the merger movement, giant corporations were formed by the combination of numerous smaller firms. The smaller companies represented nearly all the manufacturing or refining capacity of their industries. The forty largest firms in the oil-refining industry, comprising over ninety percent of the country's refining capacity and oil pipelines for its transportation, combined to form Standard Oil. In the two decades following the rise of Standard Oil, similar horizontal mergers created single dominant firms in several industries. These dominant firms included the Cottonseed Oil Trust (1884), the Linseed Oil Trust (1885), the National Lead Trust (1887), the Distillers and Cattle Feeders (1887), and the Sugar Refineries Company (1887). The trust form of organization was outlawed by court decisions. But merger activities continued to create "near" monopolies as the single corporation or holding company organization became dominant. The Diamond Match Company (1889), the American Tobacco Company (1890), the United States Rubber Company (1892), the General Electric Company (1892), and the United States Leather Company (1893) were created by the development of the modern corporation or holding company.

The height of the merger movement was reached in 1901 when 785 plants combined to form America's first billion-dollar firm, the United States Steel Corporation. The series of mergers creating the US Steel allowed it to control 65 % of the domestic blast furnace and finished steel output. This growth in concentration was typical of the first merger movement. The early mergers saw 78 of 92 large consolidations gain control of 50 % of their total industry output, and 26 secure 80 % or more.

The first major merger movement occurred during a period of rapid economic growth. The economic rationale for the large merger movement was the development of the modern corporation, with its limited liability, and the modern capital markets, which facilitated the consolidations through the absorption of the large security issues necessary to purchase firms. Nelson found that the mergers were highly correlated to the period's stock prices and industry production. However, mergers were more sensitive to stock prices. The expansion of security issues allowed financiers the financial power necessary to induce independent firms to enter large consolidations. The rationale for the first merger movement was not one of trying to preserve profits despite slackening demand and greater competitive pressures. Nor was the merger movement the result of the development of the national railroad system, which reduced geographic isolation and transportation costs. The first merger movement ended in 1904 with a depression, the onset of which coincided the Northern Securities case. Here it was held, for the first time, that antitrust laws could be used to attack mergers leading to market dominance.

A second major merger movement stirred the country from 1916 to the depression of 1929. This merger movement was only briefly interrupted by the First World War and the recession of 1921 and 1922. The approximately 12,000 mergers of the period coincided with the stock market boom of the 1920s. Although mergers greatly affected the electric and gas utility industry, market structure was not as severely concentrated by the second movement as it was by the first merger movement. Stigler (1950) concluded that mergers during this period created oligopolies, such as Bethlehem Steel and Continental Can. Mergers, primarily vertical and conglomerate in nature as opposed to the essentially horizontal mergers of the first movement, did affect competition adversely. The conglomerate product-line extensions of the 1920s were enhanced by the high-cross elasticities of demand for the merging companies' products. Antitrust laws, though not

Conference Board. An AR(1) process adequately models the quarterly mergers series, using 32 observations for the estimation period, see Table 5.1, as the partial autocorrelation (PAC) function dies after lag 1. A time series regression of mergers as a function of the components of the LEI reveals that only stock prices and the money supply are statistically significant at the 15 % level; moreover, the money supply variable has an incorrectly negative coefficient, see (5.5). An application of the Automatic Modeling Selection procedure, see (5.6), leads to only the negative money supply. Guerard reported a four-quarter lag in the relationship between mergers and stock prices from 1895 to 1954. We expect lags in the LEI to lead mergers. We use one- and two-quarter lags in the LEI data (see Table 5.2 for the cross-correlation estimate) and report in (5.8) that the one-period lagged stock price series is statistically correlated with mergers. In (5.8), (5.9), and (5.10), we report that the current and one-period lagged stock price data leads mergers. The F-statistic of (5.10) dominates the F-statistics of (5.8) and (5.9) in which we run regressions of mergers as a function of the LEI data. There is a statistically significant two-quarter lag with LEI and mergers; however, the effect is less statistically pronounced than the stock price data. An application of the Doornik and Hendry (2009a, b) Automatic Modeling Selection procedure, see (5.7), leads to a one-period lag in stock prices and four outliers. A further application of the Doornik and Hendry (2009a, b) Automatic Modeling Selection cointegration procedure, see SYS (10), leads to a one-period lag in stock prices and four outliers.

---

seriously enforced, prevented mergers from creating a single dominant firm. Merger activity diminished with the depression of 1929 and continued to decline until the 1940s.

The third merger movement began in 1940; mergers reached a significant proportion of firms in 1946 and 1947. The merger action from 1940 to 1947, although involving 7.5 % of all manufacturing and mining corporations and controlling 5 % of the total assets of the firms in those industries, was quite small compared to the merger activities of the 1920s. The mergers of the 1940s included only one merger between companies with assets exceeding 50 million dollars and none between firms with assets surpassing 100 million dollars. The corresponding figures for the mergers of the 1920s were 14 and eight, respectively. Eleven firms acquired larger firms during the mergers of the 1920s than the largest firm acquired during the 1940s merger. The mergers of the 1940s affected competition far less than did the two previous merger movements, with the exception of the food and textile industries. The acquisitions by the large firms during the 1940s rarely amounted to more than seven percent of the acquiring firms' 1939 assets or to as much as a quarter of ~ the acquiring firm's growth rate from 1940 to 1947. Approximately 5 billion dollars of assets were held by acquired or merged firms over the 1940–1947 period. Smaller firms were generally acquired by larger firms. Companies with assets exceeding 100 million dollars acquired, on average, firms with assets of less than two million dollars. The larger firms tended to engage in a greater number of acquisitions than smaller firms. The acquisitions by the larger, acquiring firms tended to involve more firms than did those acquired by smaller, acquiring firms. Mergers added relatively less to the existing size of the larger acquiring firms in the early period of the third merger movement. The relatively smaller asset growth of the larger acquiring firms is in accordance with the third merger movement's generally small effects on competition and concentration. One factor contributing to the maintenance of competition was the initiative for the mergers coming from the owners of the smaller firms. Financiers and investment bankers did not play a prominent part in the early third merger movement, but certainly have in the 1992–2011 period.

Table 5.1 Quarterly mergers, 1992–2011, autocorrelation function estimates

Sample 1 32

Autocorrelation	Partial Correlation		AC	PAC	Q-Stat	Prob
***| . |	***| . |	1	-0.430	-0.430	6.4786	0.011
. |**. |	. |* . |	2	0.289	0.129	9.5172	0.009
.**| . |	. *| . |	3	-0.297	-0.167	12.836	0.005
. |**. |	. |* . |	4	0.323	0.163	16.883	0.002
***| . |	. *| . |	5	-0.335	-0.147	21.400	0.001
. |**. |	. | . |	6	0.220	-0.027	23.434	0.001
.**| . |	. | . |	7	-0.193	-0.013	25.051	0.001
. | . |	.**| . |	8	-0.047	-0.309	25.153	0.001
. *| . |	. *| . |	9	-0.111	-0.124	25.734	0.002
. | . |	.**| . |	10	-0.028	-0.228	25.772	0.004
. |* . |	. | . |	11	0.067	0.016	26.002	0.006
. *| . |	.**| . |	12	-0.185	-0.224	27.874	0.006
. |* . |	. *| . |	13	0.121	-0.146	28.709	0.007
. | . |	. |* . |	14	0.022	0.127	28.737	0.011
. |* . |	. | . |	15	0.102	-0.057	29.399	0.014
. | . |	. |* . |	16	-0.008	0.112	29.403	0.021

Table 5.2 Quarterly mergers, 1992–2011, cross-correlation function estimates

Sample 1 32
Included observations: 32
Correlations are asymptotically consistent approximations

DDMERGERS,DLEI(-i)	DDMERGERS,DLEI(+i)	i	lag	lead
. *| . |	. *| . |	0	-0.0949	-0.0949
. *| . |	. |* . |	1	-0.1243	0.1088
. |* . |	. |***. |	2	0.1017	0.2784
.***| . |	. **| . |	3	-0.3371	-0.1761
. *| . |	. |* . |	4	-0.0897	0.1190
. | . |	. |**. |	5	-0.0390	0.1976
. |* . |	. |* . |	6	0.0523	0.1242
. **| . |	. | . |	7	-0.1949	0.0298
. |**. |	. *| . |	8	0.2065	-0.1144

If one applies the Ashley et al. (1980) transfer function causality test to the mergers and stock price series, one finds a t-value of 0.57 on the stock price series. That is, a transfer function merger model using one-period lagged stock prices as an input reduces the root mean square root relative to a random-walk with drift model, but the forecast error reduction is not statistically significant, a result reported by Guerard and McDonald (1995). Ashley (1998, 2003) and Thomakos and Guerard (2004) have reexamined the issue of post-sample periods for model validation and relative forecasting efficiency. The purpose of this case study is to present an updated and new analysis of the merger movements in the United States and the relationship between mergers, stock prices, and LEI. When we use LEI data as an input to a time series model after 2011, we must be aware that the money supply was removed as an LEI component. The money supply variable lost much of its effectiveness on economic activity in the post – 2005 period, see Levanon, Ozyildirim, Schaitkin, and Zabinska (2011). We find additional statistical correlation and regression analysis to support the historical statistical evidence that stock prices lead mergers. Stock prices are a component of the LEI; however, stock prices more directly lead mergers than the LEI. Stock prices do not lead mergers in an Ashley, Granger, and Schmalensee causality test for the 1992–2011 period.[7]

```
                    Ox Professional version 6.00

     EQ(5) Modelling dDMergers by Ordinary Least Squares (OLS)

                  Coefficient  Std.Error  t-value  t-prob Part.R^2
Constant            0.0616215    0.02423     2.54  0.0134   0.0905
dHrWeek               2.20214      1.885     1.17  0.2470   0.0206
dWkInCL             -0.406564     0.2838    -1.43  0.1568   0.0306
dMfgOrders           -1.00722     0.7157    -1.41  0.1641   0.0296
dSuppDev             0.317889     0.2910     1.09  0.2787   0.0180
dMfgNonD            0.0253327     0.2199    0.115  0.9086   0.0002
BldPerm              0.293603     0.2489     1.18  0.2425   0.0210
dSP500               0.331035     0.2130     1.55  0.1250   0.0358
dM2                  -2.53324      1.605    -1.58  0.1194   0.0369
dConExp            0.00541588     0.1708   0.0317  0.9748   0.0000

   sigma                0.107663  RSS                 0.753438773
   R^2                  0.243771  F(10,65) =     2.095 [0.037]*
   Adj.R^2              0.127429  log-likelihood          67.4866
```

The use of the Autometrics algorithm in Oxmetrics for automatic time series regressions is reported in Equation 6.

```
---------- Autometrics: dimensions of initial GUM ----------
no. of observations          76  no. of parameters          11
no. free regressors (k1)     11  no. free components (k2)    0
no. of equations              1  no. diagnostic tests        5


                 Summary of Autometrics search
initial search space       2^11  final search space        2^3
no. estimated models         93  no. terminal models         2
test form                  LR-F  target size      Default:0.05
outlier detection            no  presearch reduction      lags
backtesting                GUM0  tie-breaker                SC
diagnostics p-value        0.01  search effort        standard
time                       0.12  Autometrics version      1.5e
```

[7] Neither stock prices nor LEI passed the AGS (1980) causality test for mergers.

EQ(6) Modelling dDMergers by OLS

	Coefficient	Std.Error	t-value	t-prob	Part.R^2
Constant	0.0595716	0.01523	3.91	0.0002	0.1713
dM2	-4.34725	1.198	-3.63	0.0005	0.1511

sigma	0.106905	RSS	0.8457254
R^2	0.151143	F(1,74) =	13.18 [0.001]**
Adj.R^2	0.139672	log-likelihood	63.0958
no. of observations	76	no. of parameters	2
mean(dDMergers)	0.0267842	se(dDMergers)	0.115257

AR 1-2 test:	F(2,72) =	3.1772 [0.0476]*
ARCH 1-1 test:	F(1,74) =	0.094512 [0.7594]

The use of the lagged LEI components in the merger analysis is shown in GUM (3), and lagged stock prices are statistically significant.

GUM(3) Modelling dDMergers by OLS

	Coefficient	Std.Error	t-value	t-prob	Part.R^2
Constant	0.0293535	0.02199	1.33	0.1886	0.0373
dHrWeek	0.718022	1.935	0.371	0.7123	0.0030
dHrWeek_1	-4.76930	2.260	-2.11	0.0403	0.0883
dHrWeek_2	1.46406	1.945	0.753	0.4554	0.0122
dWkInCL	-0.268383	0.2958	-0.907	0.3690	0.0176
dWkInCL_1	0.154163	0.2948	0.523	0.6036	0.0059
dWkInCL_2	-0.0874009	0.2729	-0.320	0.7502	0.0022
dMfgOrders	-0.888739	0.7525	-1.18	0.2437	0.0294
dMfgOrders_1	0.595087	0.7798	0.763	0.4493	0.0125
dMfgOrders_2	0.397199	0.7510	0.529	0.5994	0.0060
dSuppDev	0.149083	0.2632	0.566	0.5739	0.0069
dSuppDev_1	-0.291959	0.2718	-1.07	0.2883	0.0245
dSuppDev_2	0.289764	0.2669	1.09	0.2832	0.0250
dMfgNonD	0.0375464	0.2700	0.139	0.8900	0.0004
dMfgNonD_1	-0.186103	0.2740	-0.679	0.5004	0.0099
dMfgNonD_2	-0.206999	0.2516	-0.823	0.4149	0.0145
BldPerm	-0.162543	0.2562	-0.634	0.5289	0.0087
BldPerm_1	0.231607	0.2557	0.906	0.3698	0.0175
BldPerm_2	0.166691	0.2366	0.705	0.4846	0.0107
dSP500	0.181444	0.1974	0.919	0.3627	0.0180
dSP500_1	0.374018	0.2129	1.76	0.0856	0.0629
dSP500_2	0.261796	0.2092	1.25	0.2171	0.0329
dM2	-2.36158	1.649	-1.43	0.1588	0.0427
dM2_1	-1.56727	1.631	-0.961	0.3417	0.0197
dM2_2	1.89004	1.452	1.30	0.1994	0.0355
dConExp	0.0499092	0.1528	0.327	0.7454	0.0023
dConExp_1	0.0725079	0.1710	0.424	0.6735	0.0039
dConExp_2	-0.00138546	0.1626	-0.00852	0.9932	0.0000

sigma	0.0936325	RSS	0.403284062
R^2	0.531762	F(27,46) =	1.935 [0.024]*
Adj.R^2	0.256927	log-likelihood	87.8492
no. of observations	74	no. of parameters	28
mean(dDMergers)	0.0209568	se(dDMergers)	0.10862

```
          AR 1-2 test:       F(2,44)    =   5.5118 [0.0073]**
          ARCH 1-1 test:     F(1,72)    =   10.026 [0.0023]**
           Normality test:   Chi^2(2)   =   2.2175 [0.3300]
           Hetero test:      F(54,19)   =   1.4692 [0.1790]
Chow test:       F(21,25)   =  0.71075 [0.7849] for break after 55
```

```
      ---------- Autometrics: dimensions of initial GUM ----------
      no. of observations          74  no. of parameters            28
      no. free regressors (k1)     28  no. free components (k2)      0
      no. of equations              1  no. diagnostic tests          5
```

```
                     Summary of Autometrics search
      initial search space      2^28  final search space           2^8
      no. estimated models       193  no. terminal models            4
      test form                 LR-F  target size         Default:0.05
      outlier detection           no  presearch reduction         lags
      backtesting               GUM0  tie-breaker                   SC
      diagnostics p-value       0.01  search effort           standard
      time                      0.25  Autometrics version         1.5e
```

```
                   UM(4) Modelling dDMergers by OLS

                    Coefficient  Std.Error  t-value  t-prob Part.R^2
   Constant          0.00555250    0.01152    0.482  0.6315   0.0033
   dSP500              0.291821     0.1458     2.00  0.0493   0.0541
   dSP500_1            0.515576     0.1481     3.48  0.0009   0.1475
   dSP500_2            0.204682     0.1467     1.40  0.1673   0.0271

      sigma                0.0951595  RSS                 0.633873118
      R^2                   0.264034  F(3,70) =      8.371 [0.000]**
      Adj.R^2               0.232493  log-likelihood          71.1175
      no. of observations         74  no. of parameters             4
      mean(dDMergers)      0.0209568  se(dDMergers)           0.10862

          AR 1-2 test:       F(2,68)    =   9.0433 [0.0003]**
           ARCH 1-1 test:     F(1,72)    =   2.3886 [0.1266]
          Normality test:    Chi^2(2)   =   10.374 [0.0056]**
           Hetero test:       F(6,67)   =  0.52308 [0.7888]
Chow test:       F(21,49)   =  0.66163 [0.8483] for break after 55
```

The use of the lagged LEI components in the merger analysis is shown in equation 7, EQ(7), and current and lagged stock prices are statistically significant.

```
                   EQ(7) Modelling dDMergers by OLS

                    Coefficient  Std.Error  t-value  t-prob Part.R^2
   dSP500              0.320817     0.1440     2.23  0.0290   0.0645
   dSP500_1            0.569148     0.1442     3.95  0.0002   0.1778

      sigma                0.0954224  RSS                 0.655590714
                  log-likelihood          69.871
      no. of observations         74  no. of parameters             2
      mean(dDMergers)      0.0209568  se(dDMergers)           0.10862

          AR 1-2 test:       F(2,70)    =   9.5500 [0.0002]**
           ARCH 1-1 test:     F(1,72)    =   1.9007 [0.1723]
          Normality test:    Chi^2(2)   =   10.625 [0.0049]**
           Hetero test:       F(4,69)   =  0.72801 [0.5759]
           Hetero-X test:     F(5,68)   =   1.1718 [0.3321]
           RESET23 test:      F(2,70)   = 0.058718 [0.9430]
```

```
                GUM(6) Modelling dDMergers by OLS

                 Coefficient  Std.Error  t-value  t-prob Part.R^2
dSP500              0.302460     0.1434     2.11  0.0384   0.0590
dSP500_1            0.524426     0.1462     3.59  0.0006   0.1534
dSP500_2            0.214017     0.1446     1.48  0.1433   0.0299

   sigma               0.0946435  RSS           0.635975134
                 log-likelihood          70.995
   no. of observations        74  no. of parameters           3
   mean(dDMergers)     0.0209568  se(dDMergers)         0.10862

         AR 1-2 test:        F(2,69)    =    9.0747 [0.0003]**
          ARCH 1-1 test:      F(1,72)    =    1.8115 [0.1826]
         Normality test:    Chi^2(2)   =   10.242 [0.0060]**
          Hetero test:        F(6,67)    =  0.53818 [0.7773]
Chow test:          F(21,50)  =  0.63549 [0.8713] for break after 55


                EQ(8) Modelling dDMergers by OLS

                 Coefficient  Std.Error  t-value  t-prob Part.R^2
dSP500_1            0.603289     0.1080     5.59  0.0000   0.3115
I:12                0.296530    0.07331     4.05  0.0001   0.1917
I:16                0.357167    0.07352     4.86  0.0000   0.2548
I:18                0.288096    0.07331     3.93  0.0002   0.1829
I:65               -0.179780    0.07331    -2.45  0.0167   0.0802

   sigma               0.0733045  RSS              0.370775349
                 log-likelihood          90.9588
   no. of observations        74  no. of parameters           5
   mean(dDMergers)     0.0209568  se(dDMergers)         0.10862

          AR 1-2 test:       F(2,67)    =    2.5108 [0.0888]
          ARCH 1-1 test:     F(1,72)    = 0.072879 [0.7880]
          Normality test:    Chi^2(2)   =  0.21892 [0.8963]
          Hetero test:       F(2,67)    =  0.59968 [0.5519]
          Hetero-X test:     F(2,67)    =  0.59968 [0.5519]
          RESET23 test:      F(2,67)    =   2.9589 [0.0587]

                EQ(9) Modelling dDMergers by OLS

                 Coefficient  Std.Error  t-value  t-prob Part.R^2
dDMergers_1        -0.307811     0.1105    -2.78  0.0069   0.1010
Constant           -0.0183499    0.01466    -1.25  0.2150   0.0222
dLEI                 1.41159      1.122     1.26  0.2128   0.0224
dLEI_1               1.64150      1.206     1.36  0.1779   0.0261
dLEI_2               3.29982      1.159     2.85  0.0058   0.1052

sigma              0.0963794  RSS              0.640940151
R^2                 0.255829  F(4,69) =      5.93 [0.000]**
Adj.R^2             0.212689  log-likelihood        70.7073
no. of observations       74  no. of parameters           5
mean(dDMergers)    0.0209568  se(dDMergers)         0.10862
```

```
                EQ(10) Modelling dDMergers by OLS

                   Coefficient  Std.Error  t-value  t-prob Part.R^2
dDMergers_1          -0.266408     0.1109    -2.40  0.0189   0.0742
dLEI_2                 4.16836     0.9233     4.51  0.0000   0.2206

sigma                0.0980261  RSS              0.691856946
log-likelihood         67.8789
no. of observations         74  no. of parameters          2
mean(dDMergers)      0.0209568  se(dDMergers)        0.10862

      AR 1-2 test:      F(2,70)    =   2.1666 [0.1222]
      ARCH 1-1 test:    F(1,72)    =  0.19580 [0.6595]
      Normality test:   Chi^2(2)   =   11.783 [0.0028]**
      Hetero test:      F(4,69)    =  0.37461 [0.8260]
      Hetero-X test:    F(5,68)    =  0.32981 [0.8933]
      RESET23 test:     F(2,70)    =  0.13933 [0.8702]

               SYS(10) Estimating the system by OLS

    URF equation for: dDMergers
                         Coefficient  Std.Error  t-value  t-prob
    dSP500_1                0.684853     0.1226     5.59  0.0000
    dSP500_2                0.434119     0.1293     3.36  0.0013
    dDMergers_1            -0.403148    0.09623    -4.19  0.0001
    I:12                    0.265718    0.07600     3.50  0.0009
    I:16                    0.358818    0.07591     4.73  0.0000
    I:21                  -0.0583222    0.07659   -0.762  0.4492
    I:27                  0.00245011    0.07758   0.0316  0.9749
    I:41                  -0.0283959    0.07627   -0.372  0.7109
    I:66                   0.0455124    0.07842    0.580  0.5638
    Constant       U      0.00528934   0.009625    0.550  0.5846

    sigma = 0.0751446   RSS = 0.3500962608

    URF equation for: dSP500
                        Coefficient  Std.Error  t-value  t-prob
    dSP500_1                0.197326     0.1002     1.97  0.0535
    dSP500_2               0.0479722     0.1057    0.454  0.6516
    dDMergers_1           -0.0327802    0.07869   -0.417  0.6784
    I:12                   0.0723661    0.06215     1.16  0.2487
    I:16                   0.0295421    0.06208    0.476  0.6358
    I:21                    0.179644    0.06263     2.87  0.0056
    I:27                    0.198216    0.06344     3.12  0.0027
    I:41                   -0.217843    0.06237    -3.49  0.0009
    I:66                   -0.261739    0.06413    -4.08  0.0001
    Constant       U       0.0123240   0.007871     1.57  0.1225

sigma = 0.0614508   RSS = 0.2341244214

log-likelihood         201.251  -T/2log|Omega|      405.578149
|Omega|         1.28000728e-005  log|Y'Y/T|        -9.60927303
R^2(LR)               0.809249  R^2(LM)              0.553802
no. of observations         72  no. of parameters           20

F-test on regressors except unrestricted: F(18,122) = 8.74087 [0.0000]**
F-tests on retained regressors, F(2,61) =
    dSP500_1       15.4890 [0.000]**   dSP500_2        6.16797 [0.004]**
 dDMergers_1       9.84542 [0.000]**       I:12        6.09321 [0.004]**
        I:16       12.5221 [0.000]**       I:21        6.51040 [0.003]**
        I:27       5.88011 [0.005]**       I:41        6.80811 [0.002]**
        I:66       11.5830 [0.000]**   Constant U      1.21699 [0.303]
```

Causality Testing: An Alternative Approach by Chen and Lee

The most complicated task in transfer function modeling is the identification of the transfer function form for each input series, particularly if the transfer function model includes multiple-input variables. Let us use the methodology of Liu (1999) and Chen and Lee (1990) to employ the linear transfer function (LTF) method. The LTF identification method can be used in the same manner no matter if the transfer function model has single-input or multiple-input variables. This method is more practical and easier to use than the cross correlation function (CCF) method discussed in Box and Jenkins (1976).

As in multiple regression models, a single-equation transfer function model may contain more than one input variable. Assuming that the input and output series are both stationary, the general form of a single-input transfer function model is

$$Y_t = C + \frac{\omega(B)}{\delta(B)} X_t + N_t, \quad N_t = \frac{\theta(B)}{\phi(B)} a_t, \tag{5.5}$$

where $\omega(B) = (\omega_0 + \omega_1 B + \cdots + \omega_{h-1} B^{h=1})B^b$,

$$\delta(B) = 1 - \delta_1 B - \cdots - \phi_r B^r,$$

$$\phi(B) = 1 - \phi_1 B - \cdots - \phi_p B^p,$$

and

$$\theta(B) = 1 - \theta_1 B - \cdots - \theta_q B^q.$$

The operators $\phi(B)$ and $\theta(B)$ can be in simple or multiplicative form. In the above model, N_t is referred to as the disturbance or noise of the model, and a_t is a sequence of random shocks following i.i.d. In model (5.5), the order b in the $\omega(B)$ polynomial is referred to as the *delay* of the transfer function. Box and Jenkins (1976) defined $\omega(B)$ as

$$\omega(B) = (\omega_0 - \omega_1 B - \cdots - \omega_{h-1} B^{h-1})B^b. \tag{5.6}$$

By using a positive sign in front of all ω_j coefficients, Chen and Lee (199) state that the direction of changes in Y_t will correspond to the direction of changes in X_t consistently depending on the sign of ω_j.

Similar to the stationary condition for $\phi(B)$, it is important to restrict all roots of the $\delta(B)$; it is polynomial to lie outside the unit circle. Under such an assumption, the transfer function $\omega(B)/\delta(B)$ can always be expressed in linear form as

$$V(B) = v_0 + v_1 B + v_2 B^2 + \cdots. \tag{5.7}$$

The LTF $V(B)$ has a finite number of terms. The values v_0, v_1, v_2, ... are referred to as transfer function weights (or impulse response weights) for the input series X_t. Using $V(B)$, the transfer function in (5.7) can be expressed in linear form as

$$Y_t = C + V(B)X_t + N_t. \tag{5.8}$$

Single-equation transfer function modeling also assumes a unidirectional relationship between the input and the output series, i.e., X_t may affect the present and future value of Y_t, but Y_t does not influence X_t. The same notion holds true if there are multiple-input series in the model. It is important to verify that only a unidirectional influence is present among the variables in a single-equation transfer function analysis. If a bidirectional or feedback relationship exists among the variables, inconsistent parameter estimates may occur. It is easy to extend the single-input model to multiple-input models. Assuming that we have m input variables in the system, the multiple-input transfer function model can be written as

$$Y_t = C + \frac{\omega_1(B)}{\delta_1(B)}X_{1t} + \frac{\omega_2(B)}{\delta_2(B)}X_{2t} = \cdots + \frac{\omega_m(B)}{\delta_m(B)}X_{mt} + \frac{\theta(B)}{\phi(B)}a_t, \tag{5.9}$$

where the rational transfer function $\omega_i(B)/\delta_i(B)$ for each input variable has the general form as defined in (5.9).

The identification method to be discussed in this section is applicable for both single-input and multiple-input transfer function models for notational convenience; however, the single-input model presented in (5.9) will be used here. The transfer function model identification procedure can be generally divided into three steps:

1. Estimation of the transfer function weights, v_j's
2. Determination of the model for the disturbance term N_t
3. Determination of the form of the rational polynomial $\omega(B)/\delta(B)$ that best approximates $V(B)$

The CCF is primarily used as a tool for diagnostic checking.

The rational transfer function $\omega(B)/\delta(B)$ can be approximated by an LTF $V(B)$ with a finite number of terms, say $K + 1$. Using such an approximation, model (5.10) can be expressed as

$$Y_t = C + (v_0 + v_1 B + v_2 B^2 + \cdots + v_K B^K)X_t + N_t. \tag{5.10}$$

Using the above model, the transfer function weights $v_0, v_1, v_2, \ldots, v_K$ can be easily obtained by the ordinary least squares method.

The use of the autoregressive disturbance models in the LTF method shall improve the efficiency of the transfer function eight estimates, which in turn shall improve the accuracy of the estimated disturbance $\hat{N}_j$. The values of $\hat{\phi}_1$ and $\hat{\Phi}_1$ may also provide an indication of whether regular or seasonal differencing of the input and output series is necessary. After the transfer function weights are estimated, the disturbance series can be computed using these weights where

$$\hat{N}_t = Y_t - \hat{C} - \hat{V}(B)X_t. \tag{5.11}$$

After the transfer function weights are estimated, the form of the rational transfer function $\omega(B)/\delta(B)$ can also be determined. Recall that

$$V(B) = \frac{\omega(B)}{\delta(B)} = \frac{(\omega_0 + \omega_1 B + \cdots + \omega_{h-1}B^h)B^h}{1 - \delta_1 B - \cdots - \delta_r B^r}. \tag{5.12}$$

If $\delta(B) = 1$ (i.e., $r = 0$), then $V(B) = \omega(B)$ and $V(B)$ has a cutoff pattern. On the other hand, if $\delta(B) \neq 1$ (i.e., $r \geq 1$), then $V(B)$ is an infinite series theoretically and therefore has a die-out pattern. Since $\hat{V}(B)$ is an estimate of $V(B)$, we may conclude that $\delta(B) = 1$ and $\omega(B)$ comprise only the significant terms in $\hat{V}(B)$ if $\hat{V}(B)$ has a cutoff pattern. On the other hand when $\hat{V}(B)$ has a die-out pattern, it implies that the $\delta(B)$ polynomial is not 1. In such a case, the corner table method proposed in Liu and Hanssens (1982) can be used to determine the values b, h, and r in the rational polynomial $\omega(B)/\delta(B)$.

For a set of transfer function weights v_j's, the corner table method can be used to identify the orders in the corresponding rational transfer function $\omega(B)/\delta(B)$. The method uses a table which consists of $\Delta(f, g)$ as the entry of the f-th row and g-th column, $f = 0, 1, 2, \ldots$, $g = 1, 2, 3, \ldots$, and $\Delta(f, g)$ is the determinant of a $g \times g$ matrix defined as

$$D(f, g) = \begin{bmatrix} u_f & u_{f-1} & \cdots & u_{f-g+1} \\ u_{f+1} & u_f & \cdots & u_{f-g+2} \\ \vdots & \vdots & \cdots & \vdots \\ u_{f+g-1} & u_{f+g-2} & \cdots & u_f \end{bmatrix},$$

where $u_j = v_j/v_{\max}$, $u_j = 0$ if $j < 0$, and $v_{\max}$ is the maximum value of $|v_j|$, $j = 1, 2, \ldots, K$. It can be shown that the transfer function weights v_j's have a representation $\omega(B)/\delta(B)$ with order b, h, and r if the associated table has the following pattern:

	g \ f	1	2	. . .	r-1	r	r+1	r+2	. . .
	0	0	0	. . .	0	0	0	0	. . .
	1	0	0	. . .	0	0	0	0	. . .
b	.	.	.		.	.	.	.	. . .
	.	.	.		.	.	.	.	. . .
	.	.	.		.	.	.	.	. . .
	b-1	0	0	. . .	0	0	0	0	. . .
	b	x	x	. . .	x	x	x	x	. . .
	.	.	.		.	.	.	.	. . .
h	.	.	.		.	.	.	.	. . .
	h+b-1	x	x	. . .	x	x	x	x	. . .
	h+b	*	*	. . .	*	x	0	0	. . .
	h+b+1	*	*	. . .	*	x	0	0	. . .
	.	.	.	. . .	.	.	.	.	. . .
	.	.	.	. . .	.	.	.	.	. . .
		r							

where a "0" denotes a zero value, an "x" denotes a nonzero value, and an "*" denotes an indefinite value (may or may not be zero). In the above table, the entries in the first b rows and the lower right-hand corner starting at row $h + b + 1$ (labeled as $h + b$) and column $r + 1$ are all zeros. Therefore, this table can be used to determine the values of b, h, and r, which as the corner table for the associated transfer function weights. The CCM was introduced by Tiao and Tsay (1983).

In practice the weights v_j are estimated, and the estimates $\hat{v}_j$ are subject to random errors. Consequently, one usually finds some small values in the corner table (for the zeros indicated above). However, the upper section and lower right-hand corner will show a sudden drop in values. Note that in the construction of the corner table, we have $\Delta(f, 1) = \hat{u}_f$ for the entries in the first column (i.e., when $g = 1$). Since $\hat{u}_f$ is the transfer function weight $\hat{u}_f$ normalized by $\hat{v}_{\max}$, the significance level of the values in the first column is the same as the corresponding transfer function eights estimates. For the entries in the rest of the table, one compares the absolute values of the entries with 1.0 to determine if the entries should be regarded as zeros. After a transfer function model is identified, the next step is to estimate its parameters. Representing the transfer function model as

$$Y_t = C + \frac{\omega(B)}{\delta(B)} X_t + \frac{\theta(B)}{\phi(B)} a_t, \tag{5.13}$$

the task is to estimate the vectors of parameters $\omega = [\omega_0, \omega_1, \ldots, \omega_{s-1}]'$, and $\delta = [\delta_1, \delta_2, \ldots, \delta_r]'$, $\phi = [\phi_1, \phi_2, \ldots, \phi_p]'$, and $\theta = [\theta_1, \theta_2, \ldots, \theta_q]'$. If there are several explanatory variables we will have sever ω and δ vectors. The exact ML method can be used to estimate the parameters in the transfer function model.

After a transfer function model has been identified and estimated, it is necessary to verify if the model adequately fits the data. In the same way that the sample ACF is used in the diagnostic checking of ARIMA models, the sample CCF can be used in diagnostic checking of transfer function models. The sample ACF and CCF can be conveniently combined into sample cross correlation matrices (CCM), which can be used to simplify the diagnostic checking procedure. The autocorrelation of a time series represents the correlation between the values within a series.

It is useful to note that the cross correlation at lag k is a generalization of autocorrelation at lag k since $\rho_{YX}(k) = \rho_Y(k)$ cross correlation measures not only the strength of an association but also its direction. To see the full picture of the relationship between the series Y_t and X_t, it is important to examine the cross correlations, $\rho_{YK}(k)$, for both positive and negative lags. The sequence of cross correlations $\rho_{YK}(k)$, $k = 0, \pm1, \pm2, \pm3, \ldots$ is referred to as the *CCF* for the bivariate series Y_t and X_t.

The estimate of the cross covariance at lag k, $\gamma_{YX}{}^{(k)}$ in (5.28) is provided by

$$\begin{aligned} C_{YK}(k) &= \frac{1}{n}\sum\nolimits_{t=k+1}^{n} (Y_t - \bar{Y})(X_{t-k} - \bar{X}), \quad k = 0,\ 1,\ 2,\ \ldots \\ C_{YK}(k) &= \frac{1}{n}\sum\nolimits_{t=1}^{n+k} (Y_{t-k} - \bar{Y})(X_t - \bar{X}), \quad k = 0, -1, -2,\ \ldots \end{aligned} \tag{5.14}$$

and $\bar{Y}$ and $\bar{X}$ are the sample means of Y_t and X_t series. Note that $C_{YY}(0)$ and $C_{XX}(0)$ are the estimates of σ_Y^2 and σ_X^2, respectively.

While it is workable to use CCF in diagnostic checking if only two series are considered, it is necessary to put the relevant CCFs into a matrix form to facilitate visual inspection when more than two series are involved in a study. This matrix form CCF is referred to as CCM. Assuming that $Z_t = [Y_t, X_t]'$, the CCM for the vector series Z_t are

$$\begin{array}{lcccc} \text{lag} & 0 & 1 & 2 & 3 \\ \text{CCM} & \begin{bmatrix} 1 & \rho_{YX}(0) \\ \rho_{YX}(0) & 1 \end{bmatrix} & \begin{bmatrix} \rho_{YY}(1) & \rho_{YX}(1) \\ \rho_{XY}(1) & \rho_{XX}(1) \end{bmatrix} & \begin{bmatrix} \rho_{YY}(2) & \rho_{YX}(3) \\ \rho_{XY}(2) & \rho_{XX}(2) \end{bmatrix} & \begin{bmatrix} \rho_{YY}(3) & \rho_{YX}(3) \\ \rho_{XY}(3) & \rho_{XX}(3) \end{bmatrix}. \end{array}$$

Thus the CCM contains the ACF for each series and both directions of CCFs.

When the vector series Z_t contains m time series, i.e., $Z_t = [Z_{1t}, Z_{2t}, \ldots, Z_{mt}]'$, the lag k CCM of the vector series Z_t is defined as

$$\rho(k) = \begin{bmatrix} \rho_{11}(k) & \rho_{12}(k) & \cdots & \rho_{1m}(k) \\ \rho_{21}(k) & \mathrm{P}_{22}(k) & \cdots & \mathrm{P}_{2m}(k) \\ \vdots & \vdots & \cdots & \vdots \\ & & \cdots & \\ \rho_{ml}(k) & \rho_{m2}(k) & \cdots & \rho_{mm}(k) \end{bmatrix}, \quad k = 0, 1, 2, 3, \ldots, \tag{5.15}$$

where

$$\rho_{ij}(k) = \gamma_{ij}(k) \Big/ [\gamma_{ii}(0)\gamma_{jj}(0)]^{1/2}$$

and

$$\gamma_{ij}(k) = E[(Z_{it} - \mu_i)(Z_j(t-k) - \mu_j)], \quad \mu_i = E(Z_{it}).$$

Since the cross covariance $\gamma_{ij}(k)$ can be estimated by

$$C_{ij}(k) = \frac{1}{n} \sum\nolimits_{t=k+1}^{n} (Z_{it} - \bar{Z}_i)(Z_{j(t-k)} - \bar{Z}_j), \tag{5.16}$$

the estimate of the cross correlation at lag k can be written as

$$\hat{\rho}_{ij}(k) = C_{ij}(k) \Big/ [C_{ii}(0)C_{jj}(0)]^{1/2}. \tag{5.17}$$

The (i, j)th element of the displayed lag k matrix reflects the correlation between Z_{it} and $Z_{j(t-k)}$. In this manner, the elements of the CCM and the autoregression matrices have similar interpretations.

The CCM provides an effective means to display the autocorrelations and cross correlations jointly. The autocorrelations are represented along the matrix diagonal while the cross correlations are represented by the off-diagonal elements. Interpreting the sample CCM may be difficult due to the number of entries in the matrices. Following Tiao and Box (1981), an effective summary of the correlation structure is provided by using the indicator symbols (+, −) to replace the numerical values of the elements in $\hat{\rho}(k)$ matrices, where a "+" sign is employed to indicate a value greater than $1.96/\sqrt{n}$, a "−" sign for a value less than $-1.96/\sqrt{n}$, and a "." for values in between. This device is motivated from the consideration that if the series were white noise, i.e., $Z_{it} = Z_{jt} = a_t$, then for large n, the $\rho_{ij}(k)$ would be normally distributed with mean 0 and variance n^{-1}. The standard ARIMA error diagnostics, discussed in Chapter 3, apply to multivariate or transfer function models.

The sample ACF of the residual series $\hat{a}_t$ should be examined. If $\hat{a}_t$ is indeed a white noise process, all the sample autocorrelations of the residual series should be insignificant. To verify assumption (b), the CCF between the residuals and prewhitened input series should be examined. If a_t and X_t are independent, none of the sample cross correlations should be significant. One can combine the above two steps into one step by suing sample CCM of the residuals and prewhitened input series. Assuming the independence of the residuals and the prewhitened input series, the CCMs between these two series would have insignificant values for the entire matrix over all lags as shown below:

$$\begin{bmatrix} \cdot & \cdot \\ \cdot & \cdot \end{bmatrix}$$

The diagonal elements again represent the sample autocorrelations of the $\hat{a}_t$'s and the prewhitened input series while the off-diagonal elements represent the cross correlations of these series. The dots represent insignificant correlations. If any of these correlations were significant, a "+" or "−" would appear in the relevant matrix element. Prewhitening the input series is required to correctly test for the independence of two series. Suppose that the residual series a_t is white noise but the X_t series is autocorrelated. The resulting CCF would have a pattern very similar to the ACF of the X_t series. Thus an independence test using CCF can be conducted only when each series is serially uncorrelated.

Causality Analysis of Quarterly Mergers, 1992–2011: An Application of the Chen and Lee Test

Let us consider an economic system with two variables denoted as Y_t, mergers, andCausality Analysis of Quarterly Mergers, 1992–2011... X_t, LEI or stock prices. Denoting the optimal and unbiased forecast of Y_{n+1} using the information set Ω by $\hat{Y}_{n+1}$, the conditional variance of the forecast error (which is $Y_{n+1} - \hat{Y}_{n+1}$) can be written as $\text{Var}(Y_{n+1}|\Omega)$. If the information set Ω is Y, X, or $\{Y \text{ and } X\}$ (i.e., including all data in each variable up to and including $t = n$), $\text{Var}(Y_{n+1}|\Omega)$ is the one-step-ahead forecast variance of Y_{n+1} based on Y, X, or $\{Y \text{ and } X\}$, respectively. Below are the definitions of these four possible relationships in Chen and Lee (1990):

1. Independency ($Y \wedge X$). Y and X are *independent* if and only if

$$\text{Var}(Y_{n+1}|Y) = \text{Var}(Y_{n+1}|Y,X) = \text{Var}(Y_{n+1}|Y,\ X,\ X_{n+1}) \tag{5.18}$$

and

$$\text{Var}(X_{n+1}|X) = \text{Var}(X_{n+1}|Y,X) = \text{Var}(X_{n+1}|Y,X,Y_{n+1}). \tag{5.19}$$

When two time series are independent, the one-step-ahead forecast variance of Y_{n+1} based on Y will not be reduced by including additional information on X, or including both X and concurrent information X_{n+1}. Similarly, the same relationship must also hold true for the one-step-ahead forecast variance of X_{n+1}.

Therefore when two time series are truly *independent*, no external information (including up to the forecast origin and concurrent) can improve the one-step-ahead forecast variance of Y_{n+1} or X_{n+1}.

2. Contemporaneous ($Y \leftrightarrow X$): Y and X are *contemporaneously related* if and only if

$$\mathrm{Var}(Y_{n+1}|Y) = \mathrm{Var}(Y_{n+1}|Y, X) \tag{5.20}$$

$$\mathrm{Var}(Y_{n+1}|Y,\ X) > \mathrm{Var}(Y_{n+1}|Y, X, N_{n+1}) \tag{5.21}$$

and

$$\mathrm{Var}(X_{n+1}|X) = \mathrm{Var}(X_{n+1}|Y, X) \tag{5.22}$$

$$\mathrm{Var}(X_{n+1}|Y,\ X) > \mathrm{Var}(X_{n+1}|Y, X, Y_{n+1}). \tag{5.23}$$

When two time series are *contemporaneously* related, the one-step-ahead forecast variance of Y_{n+1} based on Y will not be reduced by including additional information on X. However, when concurrent information X_{n+1} for the variable X is used, the one-step-ahead forecast variance of Y_{n+1} will be reduced. Similarly, the same relationship must also hold true for the one-step-ahead forecast variance of X_{n+1}.

3. Unidirectional ($Y \Leftarrow X$): There is a *unidirectional relationship* from X to Y if and only if

$$\mathrm{Var}(Y_{n+1}|Y) > \mathrm{Var}(Y_{n+1}|Y, X) \tag{5.24}$$

and

$$\mathrm{Var}(X_{n+1}|X) > \mathrm{Var}(X_{n+1}|Y, X). \tag{5.25}$$

When Y is *unidirectionally* influenced by X (i.e., X causes Y), the one-step-ahead forecast variance of Y_{n+1} based on Y will be reduced by including additional information on X. However, the one-step-ahead forecast variance of X_{n+1} based on X will not be reduced by including additional information on Y.

4. Feedback ($Y \Leftrightarrow X$): There is a *feedback relationship* between Y and X if and only if

$$\mathrm{Var}(Y_{n+1}|Y) > \mathrm{Var}(Y_{n+1}|Y, X) \tag{5.26}$$

and

$$\mathrm{Var}(X_{n+1}|X) > \mathrm{Var}(X_{n+1}|Y, X). \tag{5.27}$$

When Y and X have a *feedback* relationship, the one-step-ahead forecast variance of Y_{n+1} based on Y will be reduced by including additional information X, and similarly, the one-step-ahead forecast variance of X_{n+1} based on X will also be reduced by including additional information on Y.

In causality testing, our goal is to determine which dynamic relationship exists between the variables Y and X. Chen and Lee (1990) need the reader to systematically test the following five statistical hypotheses:

$$\begin{aligned}&H_1 : Y \wedge X;\\&H_2 : Y \leftrightarrow X;\\&H_3 : Y \not\Leftarrow X;\\&H_4 : Y \not\Rightarrow X;\ \text{and}\\&H_5 : Y \Leftrightarrow X.\end{aligned} \tag{5.28}$$

The hypotheses H_3 and H_4 are stated in a negative manner.

A number of time series models can be employed for causality testing (see, e.g., Sims 1972; and AGS 1980). Because VARMA models have been shown to be effective in forecasting, this class of models can also be used for causality testing (Chen and Lee 1990). A bivariate VARMA (p, q) model can be generally expressed as

$$\left(I - \phi_1 B - \cdots - \phi_p B^p\right)\begin{bmatrix} Y_t \\ X_t \end{bmatrix} = C + \left(I - \theta_1 B - \cdots = \theta_q B^q\right)\begin{bmatrix} a_{1t} \\ a_{2t} \end{bmatrix}, \tag{5.29}$$

where ϕ_i's and θ_j's are 2×2 matrices, C is a 2×1 constant vector, and $a_t = [a_{1t}, a_{2t}]'$ is a sequence of random shock vectors identically and independently distributed as a normal distribution with zero mean and covariance matrix $\sum$ with $\sum = \begin{bmatrix} \sigma_{11} & \sigma_{12} \\ \sigma_{21} & \sigma_{22} \end{bmatrix}$.

For convenience, the model in (5.29) can be rewritten as

$$\begin{bmatrix} \phi_{11}(B) & \phi_{12}(B) \\ \phi_{21}(B) & \phi_{22}(B) \end{bmatrix}\begin{bmatrix} Y_t \\ X_t \end{bmatrix} = C + \begin{bmatrix} \theta_{11}(B)\theta_{12}(B) \\ \theta_{21}(B)\theta_{22}(B) \end{bmatrix}\begin{bmatrix} a_{1t} \\ a_{2t} \end{bmatrix}, \tag{5.30}$$

where $\phi_{ij}(B) = \phi_{ij0} - \phi_{ij1}B - \phi_{ij2}B^2 - \cdots$, and $\theta_{ij}(B) = \theta_{ij0} - \theta_{ij1}B - \theta_{ij2}B^2 - \cdots$. It is important to note that $\phi_{ij0} = \theta_{ij0} = 1$ if $i = j$, and $\phi_{ij0} = \theta_{ij0} = 0$ if $i \neq j$.

Assuming that the form of the model in (5.30) is known, sufficient conditions for testing the hypotheses H_1, H_2, H_3, H_4, and H_5 using $\phi_{ij}(B)$ and $\theta_{ij}(B)$ of (5.30) are listed below:

Hypothesis	Sufficient conditions (constraints)
$H_1 : Y \wedge X$	$\phi_{12}(B) = \phi_{21}(B) = 0, \quad \theta_{12}(B) = \theta_{21}(B) = 0, \quad \sigma_{12} = \sigma_{21} = 0.$
$H_2 : Y \leftrightarrow X$	$\phi_{12}(B) = \phi_{21}(B) = 0, \quad \theta_{12}(B) = \theta_{21}(B) = 0.$
$H_3 : Y \not\Leftarrow X$	$\phi_{12}(B) = \theta_{12}(B) = 0.$
$H_4 : Y \not\Rightarrow X$	$\phi_{12}(B) = \theta_{21}(B) = 0.$
$H_5 : Y \Leftrightarrow X$	No constraints.

(53.1)

The conditions in (5.32) become necessary and sufficient conditions if the model in (5.31) is a pure vector AR or a pure vector MA model. In the above hypotheses, H_3 implies that the past X does not help to predict future Y, and H_4 implies that the past Y does not help to predict X. In both situations, we assume σ_{12} to be nonzero. However, if σ_{12} equals to zero, the hypotheses H_3, H_4, and H_5 can be tested under a more stringent condition. Therefore the following three additional hypotheses should also be considered:

Hypothesis	Sufficient conditions (Constraints)
$H_3^* : Y < \not\Leftarrow X$	$\phi_{12}(B) = \theta_{12}(B) = 0, \sigma_{12} = 0.$
$H_4^* : Y \not\Rightarrow > X$	$\phi_{21}(B) = \theta_{21}(B) = 0,$
$H_5^* : Y < \Leftrightarrow > X$	$\phi_{12} = 0.$ (5.32)

In the above hypotheses, H_3^* implies that both past and concurrent X do not help to predict Y, and H_4^* implies that both past and concurrent Y do not help to predict X. For H_5^*, it implies a "true" feedback relationship since Y and X are not contemporaneously related.

Chen and Lee (1990) proposed a decision tree approach which consists of testing a sequence of pair-wise hypotheses that are defined by each of the above relationships. This inference procedure is based on the principle that a maintained hypothesis should not be rejected unless there is sufficient evidence against it. Two procedures for identifying dynamic relationships are considered here: (1) the backward procedure and (2) the forward procedure. The backward procedure takes the position that a hypothesis should not be rejected in favor of a more restrictive one unless sufficient evidence indicates otherwise. Consequently, the statistical procedure starts from the most general hypothesis, H_5, and then examines the relative validity of competing hypotheses in an increasing order of parameter restrictions. On the other hand, the forward procedure asserts that a simpler model is preferred unless the evidence strongly suggests otherwise. Hence, the forward procedure starts its test from the most restrictive hypothesis, H_1, and moves toward less restrictive hypotheses. In both procedures, each step of the test examines one or two pairs of nested hypotheses. Chen and Lee (1990) state that the forward procedure works better (i.e., the test procedure has higher discriminating power) if the variables considered are likely to be independent or have a more restrictive relationship. On the other hand, the backward procedure works better if the variables considered are likely to have more complex relationships.

The first step of backward procedure, B1, is to examine two pairs of hypotheses: (a) H_3 versus H_5 and (b) H_4 versus H_5. This step, distinguishing the feedback relationship from unidirectional relationship, gives rise to four possible outcomes, E_1 to E_4, as follows:

E_1: H_3 is not rejected in the pair-wise test (a) and H_4 is rejected in the pair-wise test (b).
E_2: H_3 is rejected in test (a) and H_4 is not rejected in test (b).
E_3: H_3 is not rejected in test (a) and H_4 is not rejected in (b).
E_4: H_3 is rejected in test (a) and H_4 is rejected in text (b).

The outcome of E_1 implies that the past information of Y may help to predict current X, but the past X does not help to predict current Y. Hence, this outcome leads to the next pair-wise test (g), H_3^* versus H_3, where we try to detect the contemporaneous effect in the unidirectional relationship. If H_3^* is rejected in test (g), the conclusion, $Y \Rightarrow X$, is reached; otherwise the conclusion, $Y \Rightarrow> X$, would be made. Similarly, the occurrence of events E_2 and E_4, respectively, suggests a possible unidirectional relationship from X to Y and a possible feedback relationship between Y and X. Therefore, the outcome of E_2 leads to the pair-wise test (h), which helps us to choose between H_4^* and H_4. Under the outcome of E_4, it requires the test (i) which discriminates between the strong feedback hypothesis (H_5^*) and the weak feedback hypothesis (H_5). The rejection of H_4^* in test (h) implies $Y \Leftarrow X$. Otherwise, the conclusion, $Y <\Leftarrow X$, would be reached. In test (i), the rejection of H_5^* implies $Y \Leftrightarrow X$. If H_5^* is not rejected, we can conclude $Y <\Leftrightarrow> X$.

When one of the events, E_1, E_2, and E_4, occurs in sequence B1, the backward procedure stops at the end of test (g), test (h), and test (i) respectively. If neither H_3 nor H_4 is rejected (i.e., E_3 is realized), the backward procedure will move to sequence B2 where two pairs of hypotheses will be examined: (c) H_2 versus H_3 and (d) H_2 versus H_4. Again, four possible results may come out of this sequence. They are summarized as follows:

E_5: H_2 is rejected in pair-wise test (c) but is not rejected in test (d).
E_6: H_2 is not rejected in test (c) but is rejected in test (d).
E_7: H_2 is rejected in either test (c) or (d).
E_8: H_2 is rejected in both test (c) and text (d).

Since test (c) examines the possibility of $Y \Rightarrow X$ and test (d) examines that of $Y \Leftarrow X$, outcome E_5 implies that the relationship $Y \Rightarrow X$ is more probable than $Y \Leftarrow X$. Therefore, the result of event E_5 leads to test (g). A similar argument suggests that the occurrence of E_6 leads to test (h). A definitive conclusion will be reached at the end of tests (g) and (h). The rejection of H_2 in both test (c) and test (d) indicates the equal possibility of $Y \Leftarrow X$ and $Y \Rightarrow X$. Hence, the result of E_8 calls for test (f): H_2 versus H_5. If H_2 is rejected at test (f), then the possibility of the feedback relationship is established and the backward procedure moves to test (i). When H_2 is not rejected at test (f) or when event E_7 is realized, the backward procedure then proceeds to test (e), which discriminates between the independency and the contemporaneous relationship. If H_1 is rejected in test (e), the conclusion of $Y \leftrightarrow X$ is reached. Otherwise, $Y \wedge X$ will be the case.

The forward procedure, as illustrated in the previous section, begins by testing the validity of the independency hypothesis at sequence F1. The hypothesis indices, H_1 to H_5, the outcome indices, E_1 to E_8, and the pair-wise test indices, (a) to (h), are consistent. The sequence F1 considers two pairs of hypotheses testing, test (e) and test (j). If h_1 is not rejected in either test, the conclusion of $Y \wedge X$ is reached and the forward procedure stops. Otherwise, the procedure will move forward to sequence

F2, which examines the relative likelihood of the contemporaneous relationship versus the unidirectional relationship. Notice that sequence F2 is identical to sequence B2, where one of the four possible outcomes, E_5, E_6, E_7, and E_8, will emerge. Using the same argument on sequence B2, the outcomes of E_5 and E_6 lead to tests (g) and (h), respectively. A conclusion from one of the four possible unidirectional relationships can be reached as a result and the forward procedure stops. The outcome of E_7 implies $Y \leftrightarrow X$ and stops the forward procedure. However, the outcome of E_8, which rules out the case of a contemporaneous relationship, leads the forward procedure to sequence F3, which corresponds to sequence B1 in the backward procedure. Tests (a) and (b) may generate one of the four possible outcomes, E_1, E_2, E_3, and E_4. Similar to sequence B1 in the backward procedure, the outcomes of E_1 and E_2 lead to tests (g) and (h), respectively. One of the four unidirectional relationships will be detected as a result and the procedure stops. The outcome of E_4 implies a possible feedback relationship, and a further study, test (i), is needed to identify its nature. When H_5^* is rejected in test (i), we conclude $Y \Leftrightarrow X$; otherwise, we conclude $Y <\Leftrightarrow> X$. The outcome E_3 implies that Y may help to predict X and X may help to predict Y, but the nature of this dynamic relationship is not clear. Therefore, test (f) is needed. When H_2 is not rejected in test (f), the conclusion $Y \leftrightarrow X$ is reached and the procedure stops. If H_2 is rejected in test (f), the procedure moves to test (i) to determine the nature of the feedback relationship. Consequently, either $Y \Leftrightarrow X$ or $Y <\Leftrightarrow> X$ is shown to exist.

In practice, the model(s) for the time series under study is unknown. However, the order of the VARMA model for the series can be determined using the model identification procedure discussed. The test procedures are rather robust with respect to the selected model as long as the order of the model is generally correct. Corresponding to each hypothesis, the parameters of the constrained model can be estimated using the maximum likelihood estimation method. The likelihood ratio statistic is then calculated for each pair of hypotheses:

$$\mathrm{LR}(\mathrm{H}_i \text{ vs. } \mathrm{H}_j) + \mathrm{l}(\mathrm{H}_i) - \mathrm{l}(\mathrm{H}_j), \tag{5.33}$$

where $\mathrm{l}(\mathrm{H}_i) = -2^*$ (log of the maximum likelihood value under H_i). The above likelihood ratio statistic follows a χ^2-distribution with v degrees of freedom where v in each test is the difference between the number of estimated parameters under the null (the more restrictive one) and the alternative (the less restrictive one) hypotheses. A chi-square table can then be used to determine the significance of the test statistic for the tested hypotheses.

In each procedure, an a significance level will be used in conducting all pairwise tests. Note that this a level is not the Type I error probability for the overall performance of the procedures. It serves only as a cutoff point in a sequential decision procedure. The smaller the a, the higher is the probability that the more restrictive hypothesis will not be rejected. Hence, taking a smaller a is equivalent to favoring the more restrictive hypotheses (i.e., simpler relationships), and taking a larger *a* is equivalent to favoring the more complicated relationships.

The above three statistical methods investigate different aspects of a multivariate time series structure. The Sims test detects the dynamic relationship from the reduced autoregressive form, and the VARMA test examines the reduced form of a VARMA structure. The implementation of the Sims test is the easiest of the three and requires the least subjective judgement. While the literature provides a few observations on the relative performance of these three tests, Granger and Newbold (1974) pointed out that the Sims test has a tendency to generate spurious correlations. The Chen and Lee (1990) test begins with a traditional transfer function model estimate shown in Tables 5.3–5.5.

We identify two outliers in the initial merger transfer function model using LEI as the input. The estimation of the Innovational Outlier (IO, a one-time event in the time series) and Level Shift (LS, a permanent change in the time series) outliers reduces the residual standard deviations by about 20 %.[8]

LEI and stock prices are statistically associated with mergers in the Chen and Lee (1990) SCA analysis.

One sees the one and two quarter lags in the LEI in the merger transfer function model equation estimate, shown in Table 5.4

Table 5.3 Mergers, LEI, and stock price causality testing: Chen and Lee (1990) test

SUMMARY FOR UNIVARIATE TIME SERIES MODEL -- TFM1

VARIABLE	TYPE OF VARIABLE	ORIGINAL OR CENTERED	DIFFERENCING
DDMERGER	RANDOM	ORIGINAL	NONE
DLEI	RANDOM	ORIGINAL	NONE

PARAMETER LABEL	VARIABLE NAME	NUM./ DENOM.	FACTOR	ORDER	CONS- TRAINT	VALUE	STD ERROR	T VALUE
1	DLEI	NUM.	1	1	NONE	2.0222	1.0804	1.87
2	DLEI	NUM.	1	2	NONE	1.8669	1.0900	1.71
3	DDMERGER	D-AR	1	1	NONE	-.2787	.1131	-2.46

EFFECTIVE NUMBER OF OBSERVATIONS . .	73
R-SQUARE	0.294
RESIDUAL STANDARD ERROR.	0.962069E-01
(-2)*LOG LIKELIHOOD FUNCTION	-0.134658E+03
AIC.	-0.126658E+03
SIC.	-0.117496E+03

--

[8] The SCA outlier estimation using stock prices as the input series is:

One sees the one and two quarter lags in the LEI with estimated outliers in Table 5.5

Table 5.4 Summary for univariate time series model—TFM1

VARIABLE	TYPE OF VARIABLE	ORIGINAL OR CENTERED	DIFFERENCING
DDMERGER	RANDOM	ORIGINAL	NONE
DLEI	RANDOM	ORIGINAL	NONE

PARAMETER LABEL	VARIABLE NAME	NUM./ DENOM.	FACTOR	ORDER	CONS- TRAINT	VALUE	STD ERROR	T VALUE
1	DLEI	NUM.	1	1	NONE	1.8650	1.0165	1.83
2	DLEI	NUM.	1	2	NONE	1.9462	1.0362	1.88
3	DDMERGER	MA	1	1	NONE	-.6603	.1902	-3.47
4	DDMERGER	D-AR	1	1	NONE	-.8640	.1284	-6.73

EFFECTIVE NUMBER OF OBSERVATIONS . . 73
R-SQUARE 0.337
RESIDUAL STANDARD ERROR. 0.932355E-01
(-2)*LOG LIKELIHOOD FUNCTION . . . . -0.139238E+03
AIC. . . . . -0.129238E+03
SIC. . . . . -0.117786E+03

SUMMARY FOR UNIVARIATE TIME SERIES MODEL -- UTSMODEL

VARIABLE	TYPE OF VARIABLE	ORIGINAL OR CENTERED	DIFFERENCING
NS	RANDOM	ORIGINAL	NONE

PARAMETER LABEL	VARIABLE NAME	NUM./ DENOM.	FACTOR	ORDER	CONS- TRAINT	VALUE	STD ERROR	T VALUE
1	NS	MA	1	1	NONE	-.6568	.1896	-3.46
2	NS	D-AR	1	1	NONE	-.8621	.1283	-6.72

TOTAL NUMBER OF OBSERVATIONS 74
EFFECTIVE NUMBER OF OBSERVATIONS . . 73
RESIDUAL STANDARD ERROR. 0.932517E-01
--

(continued)

Table 5.4 (continued)

```
             SUMMARY FOR UNIVARIATE TIME SERIES MODEL --    TFM1
  -----------------------------------------------------------------------
              VARIABLE    TYPE OF     ORIGINAL      DIFFERENCING
                                 VARIABLE   OR CENTERED

                  DDMERGER   RANDOM      ORIGINAL      NONE

                    DLEI     RANDOM      ORIGINAL      NONE
  -----------------------------------------------------------------------

PARAMETER    VARIABLE  NUM./  FACTOR  ORDER    CONS-      VALUE       STD      T
 LABEL        NAME     DENOM.                  TRAINT                 ERROR  VALUE

1             DLEI     NUM.      1       1      NONE      1.8650     1.0165   1.83
2             DLEI     NUM.      1       2      NONE      1.9462     1.0362   1.88
3           DDMERGER    MA       1       1      NONE      -.6603      .1902  -3.47
4           DDMERGER   D-AR      1       1      NONE      -.8640      .1284  -6.73


             EFFECTIVE NUMBER OF OBSERVATIONS . .               73
             R-SQUARE . . . . . . . . . . . . . .            0.337
             RESIDUAL STANDARD ERROR. . . . . . .  0.932355E-01
             (-2)*LOG LIKELIHOOD FUNCTION . . . . -0.139238E+03
             AIC. . . . . . . . . . . . . . . . . -0.129238E+03
             SIC. . . . . . . . . .

                                   ACF RES
             NAME OF THE SERIES . . . . . . . . . .              RES
             TIME PERIOD ANALYZED . . . . . . . . .    4   TO     76
             MEAN OF THE (DIFFERENCED) SERIES . . .         -0.0074
             STANDARD DEVIATION OF THE SERIES . . .          0.0929
             T-VALUE OF MEAN (AGAINST ZERO) . . . .         -0.6789

                                AUTOCORRELATIONS

    1- 12       .02  .05  .20  .21 -.07  .10 -.02 -.12  .11 -.05  .01 -.01
    ST.E.       .12  .12  .12  .12  .13  .13  .13  .13  .13  .13  .13  .13
      Q          .0   .2  3.3  6.7  7.0  7.9  7.9  9.1 10.2 10.4 10.4 10.4

   13- 24       .12  .03  .06  .09  .03  .02  .03 -.08 -.13 -.02  .05 -.07
    ST.E.       .13  .13  .13  .13  .13  .13  .13  .13  .14  .14  .14  .14
      Q        11.8 11.9 12.2 13.0 13.1 13.1 13.2 13.8 15.6 15.6 15.8 16.3

               -1.0 -0.8 -0.6 -0.4 -0.2  0.0  0.2  0.4  0.6  0.8  1.0
                 +----+----+----+----+----+----+----+----+----+----+
                                                I
                1  0.02                   +     I     +
                2  0.05                   +     IX    +
                3  0.20                   +     IXXXXX+
                4  0.21                   +     IXXXXX+
                5 -0.07                   +   XXI     +
                6  0.10                   +     IXXX  +
                7 -0.02                   +    XI     +
                8 -0.12                   +  XXXI     +
                9  0.11                   +     IXXX  +
               10 -0.05                   +    XI     +
               11  0.01                   +     I     +
               12 -0.01                   +     I     +
               13  0.12                   +     IXXX  +
               14  0.03                  +      IX     +
               15  0.06                  +      IXX    +
               16  0.09                  +      IXX    +
               17  0.03                  +      IX     +
               18  0.02                  +      IX     +
               19  0.03                  +      IX     +
               20 -0.08                  +    XXI      +
               21 -0.13                  +   XXXI      +
               22 -0.02                  +      I      +
               23  0.05                  +      IX     +
               24 -0.07                  +    XXI      +
```

Table 5.5 Summary for univariate time series model—TFM1

VARIABLE	TYPE OF VARIABLE	ORIGINAL OR CENTERED	DIFFERENCING
DDMERGER	RANDOM	ORIGINAL	NONE
DLEI	RANDOM	ORIGINAL	NONE

PARAMETER LABEL	VARIABLE NAME	NUM./ DENOM.	FACTOR	ORDER	CONS- TRAINT	VALUE	STD ERROR	T VALUE
1	DLEI	NUM.	1	1	NONE	3.3055	.8470	3.90
2	DLEI	NUM.	1	2	NONE	1.1306	.8511	1.33
3	DDMERGER	MA	1	1	NONE	.3947	.2720	1.45
4	DDMERGER	D-AR	1	1	NONE	-.0052	.3027	-.02

SUMMARY OF OUTLIER DETECTION AND ADJUSTMENT

TIME	ESTIMATE	T-VALUE	TYPE
16	0.458	5.84	IO
34	-0.034	-4.61	LS

TOTAL NUMBER OF OBSERVATIONS. 76
EFFECTIVE NUMBER OF OBSERVATIONS. 73
RESIDUAL STANDARD ERROR (WITHOUT OUTLIER ADJUSTMENT). . 0.103580E+00
RESIDUAL STANDARD ERROR (WITH OUTLIER ADJUSTMENT) . . . 0.784829E-01

--

SUMMARY FOR UNIVARIATE TIME SERIES MODEL -- TFM1

VARIABLE	TYPE OF VARIABLE	ORIGINAL OR CENTERED	DIFFERENCING
DDMERGER	RANDOM	ORIGINAL	NONE
DSP500	RANDOM	ORIGINAL	NONE

PARAMETER LABEL	VARIABLE NAME	NUM./ DENOM.	FACTOR	ORDER	CONS- TRAINT	VALUE	STD ERROR	T VALUE
1	DSP500	NUM.	1	1	NONE	.7330	.0906	8.09
2	DDMERGER	MA	1	1	NONE	.4443	.1152	3.86
3	DDMERGER	MA	1	2	NONE	-.3205	.1151	-2.78

SUMMARY OF OUTLIER DETECTION AND ADJUSTMENT

TIME	ESTIMATE	T-VALUE	TYPE
12	0.256	3.70	IO
16	0.377	5.47	IO

TOTAL NUMBER OF OBSERVATIONS. 76
EFFECTIVE NUMBER OF OBSERVATIONS. 75
RESIDUAL STANDARD ERROR (WITHOUT OUTLIER ADJUSTMENT). . 0.865030E-01
RESIDUAL STANDARD ERROR (WITH OUTLIER ADJUSTMENT) . . . 0.690335E-01.

Let us move to a final Chen and Lee (1990) merger model estimation. The final form of the mergers and LEI analysis with the CCCF and CCM analysis is shown in Table 5.6.

Table 5.6 Summary for univariate time series model—TFM1

VARIABLE	TYPE OF VARIABLE	ORIGINAL OR CENTERED	DIFFERENCING
DDMERGER	RANDOM	ORIGINAL	NONE
DLEI	RANDOM	ORIGINAL	NONE

PARAMETER LABEL	VARIABLE NAME	NUM./ DENOM.	FACTOR	ORDER	CONS- TRAINT	VALUE	STD ERROR	T VALUE
1	DLEI	NUM.	1	1	NONE	3.3055	.8470	3.90
2	DLEI	NUM.	1	2	NONE	1.1306	.8511	1.33
3	DDMERGER	MA	1	1	NONE	.3947	.2720	1.45
4	DDMERGER	D-AR	1	1	NONE	-.0052	.3027	-.02

SUMMARY OF OUTLIER DETECTION AND ADJUSTMENT

TIME	ESTIMATE	T-VALUE	TYPE
12	0.115	3.68	TC
16	0.398	6.29	IO
18	0.183	2.97	IO
29	0.153	2.73	AO
36	-0.028	-4.56	LS
42	-0.167	-2.70	IO
67	-0.151	-2.43	IO
73	-0.148	-2.39	IO

MAXIMUM NUMBER OF OUTLIERS IS REACHED

** THE OUTLIER(S) AFTER TIME PERIOD 71 OCCURS WITHIN THE
LAST FIVE OBSERVATIONS OF THE SERIES. THE IDENTIFIED TYPE
ANS THE ESTIMATE OF THE OUTLIER(S) MAY NOT BE RELIABLE

TOTAL NUMBER OF OBSERVATIONS. 76
EFFECTIVE NUMBER OF OBSERVATIONS. 73
RESIDUAL STANDARD ERROR (WITHOUT OUTLIER ADJUSTMENT). . 0.103580E+00
RESIDUAL STANDARD ERROR (WITH OUTLIER ADJUSTMENT) . . . 0.610908E-01

--

SERIES	NAME	MEAN	STD. ERROR
1	DDMERGER	0.0268	0.1145
2	DLEI	0.0075	0.0112

NOTE: THE APPROX. STD. ERROR FOR THE ESTIMATED CORRELATIONS BELOW
IS (1/NOBE**.5) = 0.11471

SAMPLE CORRELATION MATRIX OF THE SERIES

1.00	
0.26	1.00

SUMMARIES OF CROSS CORRELATION MATRICES USING +,-,., WHERE
+ DENOTES A VALUE GREATER THAN 2/SQRT(NOBE)
- DENOTES A VALUE LESS THAN -2/SQRT(NOBE)
. DENOTES A NON-SIGNIFICANT VALUE BASED ON THE ABOVE CRITERION

(continued)

Table 5.6 (continued)

```
BEHAVIOR OF VALUES IN (I,J)TH POSITION OF CROSS CORRELATION MATRIX OVER
          ALL OUTPUTTED LAGS WHEN SERIES J LAGS SERIES I

                              1            2

                       1   .+.+.+......   ++..........
                       1   ............   ............


                       2   ............   ++.........-
                       2   ............   ............

               CROSS CORRELATION MATRICES IN TERMS OF +,-,.

                         LAGS  1 THROUGH  6

         . +      + +      . .      + .      . .      + .
         . +      . +      . .      . .      . .      . .

                         LAGS  7 THROUGH 12

         . .      . .      . .      . .      . .      . .
         . .      . .      . .      . .      . .      . -

                         LAGS 13 THROUGH 18

         . .      . .      . .      . .      . .      . .
         . .      . .      . .      . .      . .      . .

                         LAGS 19 THROUGH 24

         . .      . .      . .      . .      . .      . .
         . .      . .      . .      . .      . .      . .
                              --


                 STEPAR  VARIABLES ARE ddmerger,dLEI .  @
                     ARFITS ARE 1 to 6. rccm 1 to 6

     TIME PERIOD ANALYZED . . . . . . . . . . . . . 1  TO     76
     EFFECTIVE NUMBER OF OBSERVATIONS (NOBE). . .             76

          SERIES   NAME           MEAN       STD. ERROR

             1     DDMERGER        0.0268       0.1145
             2     DLEI            0.0075       0.0112


  NOTE: THE APPROX. STD. ERROR FOR THE ESTIMATED CORRELATIONS BELOW
                      IS (1/NOBE**.5) =    0.11471


                 SAMPLE CORRELATION MATRIX OF THE SERIES

                              1.00
                            0.26  1.00

     SUMMARIES OF CROSS CORRELATION MATRICES USING +,-,., WHERE
             + DENOTES A VALUE GREATER THAN 2/SQRT(NOBE)
             - DENOTES A VALUE LESS THAN -2/SQRT(NOBE)
     . DENOTES A NON-SIGNIFICANT VALUE BASED ON THE ABOVE CRITERION
```

(continued)

Table 5.6 (continued)

```
BEHAVIOR OF VALUES IN (I,J)TH POSITION OF CROSS CORRELATION MATRIX OVER
             ALL OUTPUTTED LAGS WHEN SERIES J LAGS SERIES I


                                        1               2


                             1   .+.+.+......   ++..........
                             1   ............   ............


                             2   ............   ++.........-
                             2   ............   ............



              CROSS CORRELATION MATRICES IN TERMS OF +,-,.

                             LAGS  1 THROUGH  6

              . +       + +       . .       + .       . .       + .
              . +       . +       . .       . .       . .       . .

                             LAGS  7 THROUGH 12

              . .       . .       . .       . .       . .       . .
              . .       . .       . .       . .       . .       . -

                             LAGS 13 THROUGH 18

              . .       . .       . .       . .       . .       . .
              . .       . .       . .       . .       . .       . .

                             LAGS 19 THROUGH 24

              . .       . .       . .       . .       . .       . .
              . .       . .       . .       . .       . .       . .

                    DETERMINANT OF S(0) = 0.146494E-05

 NOTE: S(0) IS THE SAMPLE COVARIANCE MATRIX OF W(MAXLAG+1),...,W(NOBE)


                    AUTOREGRESSIVE FITTING ON LAG(S)  1

      SUMMARIES OF CROSS CORRELATION MATRICES USING +,-,., WHERE
               + DENOTES A VALUE GREATER THAN 2/SQRT(NOBE)
               - DENOTES A VALUE LESS THAN -2/SQRT(NOBE)
       . DENOTES A NON-SIGNIFICANT VALUE BASED ON THE ABOVE CRITERION
```

(continued)

Table 5.6 (continued)

```
BEHAVIOR OF VALUES IN (I,J)TH POSITION OF CROSS CORRELATION MATRIX OVER
          ALL OUTPUTTED LAGS WHEN SERIES J LAGS SERIES I

                         1              2


               1   .+.+........   .+.....+....
               1   ............   ............


               2   ............   ............
               2   ............   ............



           CROSS CORRELATION MATRICES IN TERMS OF +,-,.

                     LAGS  1 THROUGH  6

   . .      + +      . .      + .      . .      . .
   . .      . .      . .      . .      . .      . .

                     LAGS  7 THROUGH 12

   . .      . +      . .      . .      . .      . .
   . .      . .      . .      . .      . .      . .

                     LAGS 13 THROUGH 18

   . .      . .      . .      . .      . .      . .
   . .      . .      . .      . .      . .      . .

                     LAGS 19 THROUGH 24

   . .      . .      . .      . .      . .      . .
   . .      . .      . .      . .      . .      . .


          AUTOREGRESSIVE FITTING ON LAG(S)  1  2

      SUMMARIES OF CROSS CORRELATION MATRICES USING +,-,., WHERE
               + DENOTES A VALUE GREATER THAN 2/SQRT(NOBE)
                - DENOTES A VALUE LESS THAN -2/SQRT(NOBE)
      . DENOTES A NON-SIGNIFICANT VALUE BASED ON THE ABOVE CRITERION
```

(continued)

Table 5.6 (continued)

```
BEHAVIOR OF VALUES IN (I,J)TH POSITION OF CROSS CORRELATION MATRIX OVER
            ALL OUTPUTTED LAGS WHEN SERIES J LAGS SERIES I


                                1             2


                     1  ............  .......+....
                     1  ............  ............


                     2  ............  ............
                     2  ............  ............



              CROSS CORRELATION MATRICES IN TERMS OF +,-,.

                           LAGS  1 THROUGH  6

              . .      . .      . .      . .      . .      . .
              . .      . .      . .      . .      . .      . .

                           LAGS  7 THROUGH 12

              . .      . +      . .      . .      . .      . .
              . .      . .      . .      . .      . .      . .

                           LAGS 13 THROUGH 18

              . .      . .      . .      . .      . .      . .
              . .      . .      . .      . .      . .      . .

                           LAGS 19 THROUGH 24

              . .      . .      . .      . .      . .      . .
              . .      . .      . .      . .      . .      . .


               AUTOREGRESSIVE FITTING ON LAG(S)  1  2  3

      SUMMARIES OF CROSS CORRELATION MATRICES USING +,-,., WHERE
               + DENOTES A VALUE GREATER THAN 2/SQRT(NOBE)
                - DENOTES A VALUE LESS THAN -2/SQRT(NOBE)
       . DENOTES A NON-SIGNIFICANT VALUE BASED ON THE ABOVE CRITERION
```

(continued)

Table 5.6 (continued)

```
BEHAVIOR OF VALUES IN (I,J)TH POSITION OF CROSS CORRELATION MATRIX OVER
            ALL OUTPUTTED LAGS WHEN SERIES J LAGS SERIES I

                        1              2


          1    ...+........   .......+....
          1    ............   ............


          2    ............   ............
          2    ............   ............


            CROSS CORRELATION MATRICES IN TERMS OF +,-,.

                        LAGS  1 THROUGH  6

 . .      . .      . .      + .      . .      . .
 . .      . .      . .      . .      . .      . .

                        LAGS  7 THROUGH 12

 . .      . +      . .      . .      . .      . .
 . .      . .      . .      . .      . .      . .

                        LAGS 13 THROUGH 18

 . .      . .      . .      . .      . .      . .
 . .      . .      . .      . .      . .      . .

                        LAGS 19 THROUGH 24

 . .      . .      . .      . .      . .      . .
 . .      . .      . .      . .      . .      . .


            AUTOREGRESSIVE FITTING ON LAG(S)  1  2  3  4

      SUMMARIES OF CROSS CORRELATION MATRICES USING +,-,., WHERE
                + DENOTES A VALUE GREATER THAN 2/SQRT(NOBE)
                 - DENOTES A VALUE LESS THAN -2/SQRT(NOBE)
         . DENOTES A NON-SIGNIFICANT VALUE BASED ON THE ABOVE CRITERION
```

(continued)

Table 5.6 (continued)

```
BEHAVIOR OF VALUES IN (I,J)TH POSITION OF CROSS CORRELATION MATRIX OVER
            ALL OUTPUTTED LAGS WHEN SERIES J LAGS SERIES I


                                 1              2


                       1   ......-.....   ............
                       1   ............   ............


                       2   ............   ............
                       2   ............   ............



             CROSS CORRELATION MATRICES IN TERMS OF +,-,.

                          LAGS  1 THROUGH  6

             . .      . .      . .      . .      . .      . .
             . .      . .      . .      . .      . .      . .

                          LAGS  7 THROUGH 12

             - .      . .      . .      . .      . .      . .
             . .      . .      . .      . .      . .      . .

                          LAGS 13 THROUGH 18

             . .      . .      . .      . .      . .      . .
             . .      . .      . .      . .      . .      . .

                          LAGS 19 THROUGH 24

             . .      . .      . .      . .      . .      . .
             . .      . .      . .      . .      . .      . .


            AUTOREGRESSIVE FITTING ON LAG(S)  1  2  3  4  5

      SUMMARIES OF CROSS CORRELATION MATRICES USING +,-,., WHERE
               + DENOTES A VALUE GREATER THAN 2/SQRT(NOBE)
                - DENOTES A VALUE LESS THAN -2/SQRT(NOBE)
        . DENOTES A NON-SIGNIFICANT VALUE BASED ON THE ABOVE CRITERION
```

(continued)

Table 5.6 (continued)

```
BEHAVIOR OF VALUES IN (I,J)TH POSITION OF CROSS CORRELATION MATRIX OVER
            ALL OUTPUTTED LAGS WHEN SERIES J LAGS SERIES I

                            1              2

                1    ......-.....   ............
                1    ............   ............


                2    ............   ............
                2    ............   ............


             CROSS CORRELATION MATRICES IN TERMS OF +,-,.

                          LAGS  1 THROUGH  6

             . .      . .      . .      . .      . .      . .
             . .      . .      . .      . .      . .      . .

                          LAGS  7 THROUGH 12

             - .      . .      . .      . .      . .      . .
             . .      . .      . .      . .      . .      . .

                          LAGS 13 THROUGH 18

             . .      . .      . .      . .      . .      . .
             . .      . .      . .      . .      . .      . .

                          LAGS 19 THROUGH 24

             . .      . .      . .      . .      . .      . .
             . .      . .      . .      . .      . .      . .


          AUTOREGRESSIVE FITTING ON LAG(S)  1  2  3  4  5  6

      SUMMARIES OF CROSS CORRELATION MATRICES USING +,-,., WHERE
               + DENOTES A VALUE GREATER THAN 2/SQRT(NOBE)
                - DENOTES A VALUE LESS THAN -2/SQRT(NOBE)
       . DENOTES A NON-SIGNIFICANT VALUE BASED ON THE ABOVE CRITERION
```

(continued)

Table 5.6 (continued)

```
BEHAVIOR OF VALUES IN (I,J)TH POSITION OF CROSS CORRELATION MATRIX OVER
            ALL OUTPUTTED LAGS WHEN SERIES J LAGS SERIES I

                    1               2

            1  ......-.....   ............
            1  ........-...   ............

            2  ............   ............
            2  ............   ............

         CROSS CORRELATION MATRICES IN TERMS OF +,-,.

                    LAGS  1 THROUGH  6

      . .     . .     . .     . .     . .     . .
      . .     . .     . .     . .     . .     . .

                    LAGS  7 THROUGH 12

      - .     . .     . .     . .     . .     . .
      . .     . .     . .     . .     . .     . .

                    LAGS 13 THROUGH 18

      . .     . .     . .     . .     . .     . .
      . .     . .     . .     . .     . .     . .

                    LAGS 19 THROUGH 24

      . .     . .     - .     . .     . .     . .
      . .     . .     . .     . .     . .     . .
```

(continued)

Table 5.6 (continued)

```
========== STEPWISE AUTOREGRESSION SUMMARY ==========

-------------------------------------------------------------------
    I RESIDUAL I EIGENVAL.I CHI-SQ  I           I SIGNIFICANCE
LAG I VARIANCESI OF SIGMA I  TEST   I   AIC     I OF PARTIAL AR COEFF.
----+----------+----------+---------+----------+---------------------
  1 I .105E-01 I .981E-04 I   23.68 I  -13.684 I - +
    I .101E-03 I .105E-01 I         I          I . +
----+----------+----------+---------+----------+---------------------
  2 I .896E-02 I .974E-04 I   10.42 I  -13.741 I . +
    I .998E-04 I .897E-02 I         I          I . .
----+----------+----------+---------+----------+---------------------
  3 I .870E-02 I .915E-04 I    5.78 I  -13.728 I . .
    I .944E-04 I .870E-02 I         I          I . .
----+----------+----------+---------+----------+---------------------
  4 I .737E-02 I .904E-04 I   10.74 I  -13.800 I + .
    I .941E-04 I .737E-02 I         I          I . .
----+----------+----------+---------+----------+---------------------
  5 I .735E-02 I .901E-04 I     .29 I  -13.700 I . .
    I .939E-04 I .736E-02 I         I          I . .
----+----------+----------+---------+----------+---------------------
  6 I .726E-02 I .896E-04 I    1.05 I  -13.613 I . .
    I .931E-04 I .726E-02 I         I          I . .
-------------------------------------------------------------------

NOTE:  CHI-SQUARED CRITICAL VALUES WITH   4  DEGREES OF FREEDOM ARE

                5 PERCENT:   9.5     1 PERCENT:  13.3

NOTE:  THE PARTIAL AUTOREGRESSION COEFFICIENT MATRIX FOR LAG L IS THE
       ESTIMATED PHI(L) FROM THE FIT WHERE THE MAXIMUM LAG USED IS L
       (I.E. THE LAST COEFFICIENT MATRIX). THE ELEMENTS ARE
       STANDARDIZED BY DIVIDING EACH BY ITS STANDARD ERROR.
                              --

         MTSMODEL ARMA11. SERIES ARE ddmerger,dLEI. @
          MODEL IS (1-PHI*B)SERIES=C+(1-TH1*B)NOISE.


       SUMMARY FOR MULTIVARIATE ARMA MODEL --  ARMA11

              VARIABLE    DIFFERENCING

                    DDMERGER
                      DLEI

    PARAMETER     FACTOR      ORDER    CONSTRAINT

       1        C   CONSTANT     0         CC
      2      PHI    REG AR      1        CPHI
      3      TH1    REG MA      1        CTH1
                              --

   CAUSALTEST MODEL ARMA11. OUTPUT PRINT(CORR). alpha .01
```

(continued)

Table 5.6 (continued)

```
                 SUMMARY OF THE TIME SERIES

SERIES     NAME            MEAN         STD DEV    DIFFERENCE ORDER(S)

         1      DDMERGER          0.0268          0.1145
         2      DLEI              0.0075          0.0112

                    -----------------------
                    ERROR COVARIANCE MATRIX
                    -----------------------

                          1           2
                   1      .011543
              2      .000306      .000136

                ITERATIONS TERMINATED DUE TO:
     CHANGE IN  (-2*LOG LIKELIHOOD)/NOBE  .LE. 0.100E-03
             TOTAL NUMBER OF ITERATIONS IS   10

   MODEL SUMMARY WITH MAXIMUM LIKELIHOOD PARAMETER ESTIMATES

          ----- CONSTANT VECTOR (STD ERROR) -----

                    -0.045  (    0.024  )
                     0.004  (    0.002  )

                    ----- PHI MATRICES -----

     ESTIMATES OF    PHI(  1 ) MATRIX AND SIGNIFICANCE

                    -.280   10.149       . +
                     .004     .472       . +

                       STANDARD ERRORS

                        .243    3.027
                        .017     .230

                   ----- THETA MATRICES -----

     ESTIMATES OF  THETA(  1 ) MATRIX AND SIGNIFICANCE

                    -.050    8.230       . +
                    -.011     .073       . .

                       STANDARD ERRORS

                        .263    3.218
                        .018     .246


                    -----------------------
                    ERROR COVARIANCE MATRIX
                    -----------------------

                        1           2
              1      .008643
              2      .000143      .000100
```

(continued)

Table 5.6 (continued)

SUMMARY OF FINAL PARAMETER ESTIMATES AND THEIR STANDARD ERRORS

PARAMETER NUMBER	PARAMETER DESCRIPTION	FINAL ESTIMATE	ESTIMATED STD. ERROR
1	CONSTANT(1)	-0.045279	0.023810
2	CONSTANT(2)	0.003729	0.001855
3	AUTOREGRESSIVE (1, 1, 1)	-0.280315	0.243195
4	AUTOREGRESSIVE (1, 1, 2)	10.149337	3.027162
5	AUTOREGRESSIVE (1, 2, 1)	0.003976	0.016524
6	AUTOREGRESSIVE (1, 2, 2)	0.471517	0.229912
7	MOVING AVERAGE (1, 1, 1)	-0.049937	0.262984
8	MOVING AVERAGE (1, 1, 2)	8.230138	3.217553
9	MOVING AVERAGE (1, 2, 1)	-0.010919	0.018313
10	MOVING AVERAGE (1, 2, 2)	0.073078	0.245583

CORRELATION MATRIX OF THE PARAMETERS

	1	2	3	4	5	6	7	8	9	10
1	1.00									
2	-.19	1.00								
3	.24	.35	1.00							
4	-.79	-.07	-.54	1.00						
5	.08	.31	.36	-.20	1.00					
6	.01	-.79	-.45	.12	-.59	1.00				
7	.20	.33	.90	-.47	.36	-.43	1.00			
8	-.75	-.05	-.51	.94	-.19	.10	-.47	1.00		
9	-.07	.22	.10	.04	.82	-.45	.15	.04	1.00	
10	-.17	-.70	-.45	.31	-.57	.91	-.43	.30	-.43	1.00

```
THE RESIDUAL COVARIANCE MATRIX IS SET TO FULL MATRIX
ALL ELEMENTS IN THE MATRIX PARAMETERS ARE ALLOWED TO BE ESTIMATED
-2*(LOG LIKELIHOOD AT FINAL ESTIMATES UNDER H5 ) IS  -0.89830741E+03

THE RESIDUAL COVARIANCE MATRIX IS SET TO DIAGONAL MATRIX
ALL ELEMENTS IN THE MATRIX PARAMETERS ARE ALLOWED TO BE ESTIMATED
-2*(LOG LIKELIHOOD AT FINAL ESTIMATES UNDER H5*) IS  -0.89665939E+03

THE RESIDUAL COVARIANCE MATRIX IS SET TO FULL MATRIX
THE (2,1)TH ELEMENTS IN THE MATRIX PARAMETERS ARE SET TO ZERO
-2*(LOG LIKELIHOOD AT FINAL ESTIMATES UNDER H4 ) IS  -0.89659059E+03

THE RESIDUAL COVARIANCE MATRIX IS SET TO DIAGONAL MATRIX
THE (2,1)TH ELEMENTS IN THE MATRIX PARAMETERS ARE SET TO ZERO
-2*(LOG LIKELIHOOD AT FINAL ESTIMATES UNDER H4*) IS  -0.89537550E+03

THE RESIDUAL COVARIANCE MATRIX IS SET TO FULL MATRIX
THE (1,2)TH ELEMENTS IN THE MATRIX PARAMETERS ARE SET TO ZERO
-2*(LOG LIKELIHOOD AT FINAL ESTIMATES UNDER H3 ) IS  -0.87864498E+03

THE RESIDUAL COVARIANCE MATRIX IS SET TO DIAGONAL MATRIX
THE (1,2)TH ELEMENTS IN THE MATRIX PARAMETERS ARE SET TO ZERO
-2*(LOG LIKELIHOOD AT FINAL ESTIMATES UNDER H3*) IS  -0.87714380E+03

THE RESIDUAL COVARIANCE MATRIX IS SET TO FULL MATRIX
THE (2,1)TH ELEMENTS IN THE MATRIX PARAMETERS ARE SET TO ZERO
THE (1,2)TH ELEMENTS IN THE MATRIX PARAMETERS ARE SET TO ZERO
-2*(LOG LIKELIHOOD AT FINAL ESTIMATES UNDER H2 ) IS  -0.87754552E+03

THE RESIDUAL COVARIANCE MATRIX IS SET TO DIAGONAL MATRIX
THE (2,1)TH ELEMENTS IN THE MATRIX PARAMETERS ARE SET TO ZERO
THE (1,2)TH ELEMENTS IN THE MATRIX PARAMETERS ARE SET TO ZERO
-2*(LOG LIKELIHOOD AT FINAL ESTIMATES UNDER H1 ) IS  -0.87634247E+03

RESULT BASED ON THE BACKWARD PROCEDURE ( Y:DDMERGER,  X: DLEI     )
        DDMERGER <<= DLEI     (Y IS STRONGLY CAUSED BY X)

RESULT BASED ON THE FORWARD PROCEDURE  ( Y:DDMERGER,  X: DLEI     )
        DDMERGER <<= DLEI     (Y IS STRONGLY CAUSED BY X)
```

The Chen and Lee (1990) test finds that LEI strongly cause mergers during the 1992–2011 period. Moreover, the Chen and Lee (1990) test finds that stock prices cause mergers during the 1992–2011 period.[9]

Money Supply and Stock Prices, 1967–2011

We examine the causal relationship between the money supply (M1P) and stock prices, as measured by the S&P 500 during the 1967.01–2011.04 period. Thomakos and Guerard (2004) and Ashley (2004) found that the money supply passed the AGS (1980) causality test and the Ashley post-sample criteria test (2004). We obtain M1P and S&P 500 monthly data from the St. Louis Federal Reserve economic database (FRED).[10] Both series have a difference in the logarithmic process; i.e., the series are dlog-transformed. We use SCA and the Chen and Lee (1990) test for the money supply and stock returns series. There is a four-month lag in the (positive) effect of the money supply on stock prices (and returns), see Table 5.7.

Table 5.7 The money supply and stock prices, 1967–2011

PARAMETER LABEL	VARIABLE NAME	NUM./ DENOM.	FACTOR	ORDER	CONS- TRAINT	VALUE	STD ERROR	T VALUE
1	MSIM1P	NUM.	1	1	NONE	-.5574	.2845	-1.96
2	MSIM1P	NUM.	1	2	NONE	.0890	.2643	.34
3	MSIM1P	NUM.	1	3	NONE	.1484	.2641	.56
4	MSIM1P	NUM.	1	4	NONE	1.0079	.2837	3.55
5	SP500	D-AR	1	1	NONE	.2728	.0414	6.58

EFFECTIVE NUMBER OF OBSERVATIONS . . 540

R-SQUARE 0.081

RESIDUAL STANDARD ERROR. 0.357160E-01

The transfer function residuals are white noise (random), as illustrated by the autocorrelation function of the residuals (ACF RES).

(continued)

[9] Had one modeled stock prices and mergers for the 1979–2011 period, one finds only a contemporaneous relationship and no strong causality findings.

[10] We use M1P, a variation on M1, rather than M3, that was used in the earlier studies because M3 was discontinued in the FRED database.

Table 5.7 (continued)

```
                    ACF RES

      NAME OF THE SERIES..........          RES
  MEAN OF THE (DIFFERENCED) SERIES...     0.0011
  STANDARD DEVIATION OF THE SERIES...     0.0355
   T-VALUE OF MEAN (AGAINST ZERO)....     0.6960

              AUTOCORRELATIONS

 1- 12  -.01 -.00 .04 .01 .08 -.07 -.04 .05 -.01 -.02 .06 -.02
  ST.E.  .04 .04 .04 .04 .04 .04 .04 .04 .04 .04 .04 .04
  Q       .0  .0  .8  .9 4.6 7.4 8.1 9.7 9.7 10.0 11.9 12.1

13- 24  .01 -.04 -.04 .07 -.05 .03 -.00 -.06 -.08 -.01 -.03 -.03
  ST.E.  .04 .04 .04 .04 .04 .04 .04 .04 .04 .04 .04 .04
Q    12.1 12.9 13.7 16.3 17.6 18.0 18.0 19.9 23.3 23.3 23.8 24.3

      -1.0 -0.8 -0.6 -0.4 -0.2 0.0 0.2 0.4 0.6 0.8 1.0
                                 I
           1 -0.01              + I +
           2  0.00              + I +
           3  0.04              + IX+
           4  0.01              + I +
           5  0.08              + IXX
           6 -0.07              XXI +
           7 -0.04              +XI +
           8  0.05              + IX+
           9 -0.01              + I +
          10 -0.02              +XI +
          11  0.06              + IX+
          12 -0.02              + I +
          13  0.01              + I +
          14 -0.04              +XI +
          15 -0.04              +XI +
          16  0.07              + IXX
          17 -0.05              +XI +
          18  0.03              + IX+
          19  0.00              + I +
          20 -0.06              +XI +
          21 -0.08              XXI +
          22 -0.01              + I +
          23 -0.03              +XI +
          24 -0.03              +XI +
```

Table 5.8 Summary for univariate time series model—TFM1

VARIABLE	TYPE OF VARIABLE	ORIGINAL OR CENTERED	DIFFERENCING
SP500	RANDOM	ORIGINAL	NONE
MSIM1P	RANDOM	ORIGINAL	NONE

PARAMETER LABEL	VARIABLE NAME	NUM./ DENOM.	FACTOR	ORDER	CONS- TRAINT	VALUE	STD ERROR	T VALUE
1	MSIM1P	NUM.	1	1	NONE	.0180	.2289	.08
2	MSIM1P	NUM.	1	2	NONE	.0548	.2140	.26
3	MSIM1P	NUM.	1	3	NONE	.0884	.2137	.41
4	MSIM1P	NUM.	1	4	NONE	.6230	.2282	2.73
5	SP500	MA	1	1	NONE	-.2144	.0429	-5.00

SUMMARY OF OUTLIER DETECTION AND ADJUSTMENT

TIME	ESTIMATE	T-VALUE	TYPE
40	-0.120	-4.33	AO
82	-0.074	-3.21	TC
90	-0.097	-4.15	TC
96	0.107	4.62	TC
108	0.087	3.06	IO
158	-0.105	-3.67	IO
176	-0.090	-3.23	AO
188	0.104	4.50	TC
249	-0.100	-4.35	TC
283	-0.086	-3.03	IO
289	0.111	3.88	IO
364	0.082	3.54	TC
379	-0.087	-3.04	IO
382	0.080	3.46	TC
410	-0.090	-3.15	IO
416	-0.123	-4.43	AO
426	-0.110	-3.95	AO
501	-0.219	-7.70	IO
507	0.093	4.04	TC
535	-0.123	-4.33	IO

TOTAL NUMBER OF OBSERVATIONS............. 545

EFFECTIVE NUMBER OF OBSERVATIONS........... 541

RESIDUAL STANDARD ERROR (WITHOUT OUTLIER ADJUSTMENT).. 0.358532E-01

RESIDUAL STANDARD ERROR (WITH OUTLIER ADJUSTMENT)... 0.284839E-01

We find significant outliers in the money supply and stock returns series estimates, see Table 5.8.

The estimation of outliers reduces the residual standard error by approximately 20 %.

However, the Chen and Lee (1990) test does not report that the money supply causes stock prices,

```
RESULT BASED ON THE BACKWARD PROCEDURE ( Y:SP500, X: MSIM1P)
SP500 =>> MSIM1P (Y STRONGLY CAUSES X)
RESULT BASED ON THE FORWARD PROCEDURE ( Y:SP500, X: MSIM1P)
SP500 ^ MSIM1P (Y IS INDEPENDENT OF X)
```

but rather that stock prices (returns) cause the money supply and that stock prices are independent of the money supply.

In this chapter, we fit univariate and bivariate time series models in the tradition of Box and Jenkins (1976) and Granger and Newbold (1977) and apply traditional Granger causality testing following the Ashley et al. (1980) methodology and the Vector Autoregressive Models (VAR) and Chen and Lee (1990) VARMA causality test. We test two series for causality: (1) stock prices and mergers and (2) the money supply and stock prices. We find mixed results on Granger causality testing models.

References

Ashley, R.A. 2003. "Statistically Significant Forecasting Improvements: How Much Out-of-sample Data is Likely Necessary?" *International Journal of Forecasting* 19, 229-240.

Ashley, R. 1998. "A New Technique for Postsample Model Selection and Validation." *Journal of Economic Dynamics and Control* 22, 647-665.

Ashley, R. A., C. W. J., Granger, and R. L. Schmalensee. 1980. "Advertising and Aggregate Consumption: An Analysis of Causality". *Econometrica* 48, 1149-1168.

Box, G.E.P. and Jenkins, G.M. 1976. *Time Series Analysis Forecasting and Control*. San Francisco: Holden-Day.

Butters, J. K., J. Lintner, and W. L. Cary. 1951. *Corporate Mergers*. Boston: Division of Research, Graduate School of Business, Harvard University Press.

Chen, C. and Lee, C.J. (1990). "A VARMA Test on the Gibson Paradox." *Review of Economics and Statistics 72*: 96-107.

Clements, M. P. and D. F. Hendry. 1998. *Forecasting Economic Time Series*. Cambridge: Cambridge University Press.

Doornik, J. A. and D. F. Hendry. *Empirical Econometric Modelling*. Timberlake Consultants, Ltd.

Doornik, J. A. and D. F. Hendry. *Modelling Dynamic Systems*. Timberlake Consultants, Ltd.

Feige, E.L. and D. K. Pearce, 1976. "Economically Rational Expectations: Are Innovations in the Rate of Inflation Independent of Innovations in Measures of Monetary and Fiscal Policy?" *Journal of Political Economy 84*: 499-522.

Granger, C.W.J. 1969. "Investigating Causal Relations by Econometric Models and Cross-Spectral Methods." *Econometrica 37*: 424-438.

Granger, C.W.J. and P. Newbold. 1974. "Spurious Regressions in Econometrics." *Journal of Econometrics 1*: 111-120.

Granger, C.W.J. and P. Newbold. 1977. *Forecasting Economic Time Series*. New York: Academic Press.

Guerard, J.B., Jr. 1985. "Mergers, Stock Prices, and Industrial Production: An Empirical Test of the Nelson Hypothesis", in O.D. Anderson, Ed., *Time Series Analysis: Theory and Practice* 7, Amsterdam: Elsevier, 239-248.

Golbe, D. L. and L. J. White. 1993. "Catch a Wave: The Time Series Behavior of Mergers". *The Review of Economics and Statistics* 75, 493-499.

Guerard, J. B., Jr. and J. B. McDonald. 1995. "Mergers in the United States: A Case Study in Robust Time Series", in C. F. Lee., Ed., *Advances in Financial Planning and Forecasting*, Greenwich, CT: JAI Press, 95–107.

Guerard, John B., Jr. 2004. "The Forecasting Effectiveness of the U.S. Leading Economic Indicators: Further Evidence and Initial G7 Results." In P. Dua, Ed., *Business Cycles and Economic Growth: An AnalysisUsing Leading Indicators*. New York: Oxford University Press, 174-187.

Hamilton, J.D. 1994. *Time Series Analysis*, Princeton, Princeton University Press.

Hoover, K.D. 2001. *Causality in Macroeconomics*. Cambridge: Cambridge University Press.

Lee, C.J. and E. Petruzzi, E. 1986. "The Gibson Paradox and Monetary Standard." *The Review of Economics and Statistics 68*: 189-196.

Levanon G., A. Ozyildirim, B. Schaitkin, and J. Zabinska. 2011. "Comprehensive Benchmark Revisions for the conference Board Leading Economic Index for the United States." The Conference Board. EPWP # 1–06. December 2011.

Liu, L.M. 1999. *Forecasting and Time Series Analysis Using the SCA Statistical System: Volume 2*. Chicago: Scientific Computing Associates Corp.

Lutkepohl, H. 1993. *Introduction to Multiple Time Series Analysis*. 2nd edition. Berlin: Springer-Verlag.

McCracken, M. 2000. "Robust Out-of-Sample Inference." *Journal of Econometrics 39*, 195-223

Melicher, R., W. J. Ledolter, and L. J. D'Antonio. 1983. "A Time Series Analysis of Aggregate Merger Activity". *The Review of Economics and Statistics* 65, 423-430.

Nelson, C.R. and G.B. Schwertz, G.B. 1982. "Tests for Predictive Relationships Between Time Series Variables: A Monte Carlo Investigation." *Journal of the American Statistical Association 77*: 11-18.

Nelson, R. L. 1959. *Merger Movements in American Industry, 1895–1956*. Princeton: Princeton University Press.

Nelson, R. L. 1966. "Business Cycle Factors in the Choice between Internal and External Growth", in W.W. Alberts and J. E. Segall, Eds, *The Corporate Merger*. Chicago: The University of Chicago Press, 52-66.

Pierce, D.A. and L. D. Haugh, L.D. 1977. "Causality in Temporal Systems: Characterizations and Survey," *Journal of Econometrics 5*: 265-293.

Sargent, T.J. 1973. "Interest Rates and Prices in the Long Run." *Journal of Money, Credit and Banking*: 385-463.

Shiller, R.J. and Siegel, J.J. 1997. "The Gibson Paradox and Historical Movement in Real Interest Rates." *Journal of Political Economy 85*: 891-907.

Sims, C.A. 1972. "Money, Income and Causality," *American Economic Review 62*: 540-552.

Sims, C.A. 1980. "Macro-Economics and Reality." *Econometrica 48*: 1-48.

Stigler, G. J. 1950. Monopoly and Oligopoly by Merger. *American Economic Review* 40, 23-34.

Tiao, G.C. and G.E. P. Box, G.E.P. 1981. "Modeling Multiple Time Series with Applications." *Journal of American Statistical Association 76*: 802-816.

Tiao, GC. and R.S. Tsay. 1983, "Multiple Time Series Modeling and Extended Sample Cross Correlations," *Journal of Business and Economic Statistics* 1: 43–56.

Thomakos, D. and J. Guerard. 2004. "Naïve, ARIMA, Transfer Function, and VAR Models: A Comparison of Forecasting Performance." *The International Journal of Forecasting* 20, 53-67.

Wicksell, K. 1907. "The Influence of the Rate of Interest on Prices." *Economic Journal 17*: 213-220.

Zellner, A. and Palm, F. 1974. "Time Series Analysis and Simultaneous Equation Econometric Models." *Journal of Econometrics* 2: 17-54.

Zarnowitz, V. 1992. *Business Cycles: Theory, History, Indicators, and Forecasting*. Chicago: The University of Chicago Press.

Chapter 6
A Case Study of Portfolio Construction Using the USER Data and the Barra Aegis System

In this chapter, we estimate a set of monthly regression models to create monthly expected returns and demonstrate the effectiveness of the Barra Aegis system. The Aegis system creates and tests investment management strategies, producing portfolios and attributing portfolio returns according to the Barra multifactor risk model. This chapter draws heavily on Rudd and Clasing (1982) and Grinold and Kahn (2000) for complete descriptions of the Barra system. We find support with the Barra Aegis for the composite modeling, the United States Expected Returns (USER), developed and estimated in Chapter 4, using fundamental, expectations, and momentum-based data for the US equities during the December 1979–December 2009 period. To measure risk, one can vary the period of volatility calculation, such as using 5 years of monthly data in calculating the covariance matrix, as was done in Bloch, Guerard, Markowitz, Todd, and Xu (1993), or 1 year of daily returns to calculate a covariance matrix, as was done in Guerard, Takano, and Yamane (1993), or 2–5 years of data to calculate factor returns as in the Barra system, discussed in Menchero, Morozov, and Shepard (2010). The Capital Asset Pricing Model, the CAPM, as put forth by Sharpe (1964), Lintner (1965a), and Mossin (1966), holds that the return to a security is a function of the security beta:

$$R_{jt} = R_{\mathrm{F}} + \beta_j[E(R_{\mathrm{Mt}}) - R_F] + e_{jt}, \tag{6.1}$$

J.B. Guerard, Jr., *Introduction to Financial Forecasting in Investment Analysis*,

DOI 10.1007/978-1-4614-5239-3_6,

where R_{jt} is expected security return at time t; $E\ (R_{Mt})$, expected return on the market at time t; R_F, risk-free rate; β_j, security beta, a random regression coefficient; and e_{jt}, randomly distributed error term.[1]

Let us estimate beta coefficients to be used in the CAPM to determine the rate of return on equity. One can fit a regression line of monthly holding period returns (HPRs) against the excess returns of an index such as the value-weighted Center for Research in Security Prices (CRSP) index, which is an index of all publicly traded stocks. Most stock betas are estimated using 5 years of monthly data, some sixty observations, although one can use almost any number of observations.[2] One generally needs at least thirty observations for normality of residuals to occur. One can use the Standard & Poor's 500 Index, or the Dow Jones Industrial Index (DJIA), or many other stock indexes.

Empirical tests of the CAPM often resulted in unsatisfactory results. That is, the average estimated market risk premium was too small, relative to the theoretical market risk premium and the average estimated risk-free rate exceeded the known risk-free rate. Thus low-beta stocks appeared to earn more than was expected and high-beta stocks appeared to earn less than was expected (Black, Jensen, and Scholes (1972)). The equity world appeared more risk-neutral than one would have expected during the 1931–1965 period. There could be many issues with estimating betas using ordinary least squares. Roll (1969, 1977) and Sharpe (1971) identified and tested several issues with beta estimations. Bill Sharpe estimated characteristic lines, the line of stock or mutual fund return versus the market return, using ordinary least squares (OLS) and the mean absolute deviation (MAD) for the 30 stocks of the Dow Jones Industrial Average stocks versus the Standard and Poor's 425 Index (S&P 425) for the 1965–1970 period and 30 randomly selected mutual funds over the 1964–1970 period versus the S&P 425. Sharpe found little difference in the OLS and MAD betas, and concluded that the MAD estimation gains may be "relatively modest."

---

[1] The CAPM beta, its measure of systematic risk, from the Capital Market Line equilibrium condition, in an alternative formulation:

$$\beta_j = \frac{\mathrm{Cov}(R_j, R_\mathrm{M})}{\mathrm{Var}(R_\mathrm{M})} \tag{6.2}$$

$$\begin{aligned} E(R_j) &= R_\mathrm{F} + \left[\frac{E(R_\mathrm{M}) - R_\mathrm{F}}{\sigma_\mathrm{M}^2}\right] \mathrm{Cov}(R_j, R_\mathrm{M}) \\ &= R_\mathrm{F} + [E(R_\mathrm{M}) - R_\mathrm{F}] \frac{\mathrm{Cov}(R_j, R_\mathrm{M})}{\mathrm{Var}(R_\mathrm{M})} \end{aligned}$$

$$E(R_j) = R_\mathrm{F} + [E(R_\mathrm{M}) - R_\mathrm{F}]\beta_j. \tag{6.3}$$

Equation (6.3) defines the Security Market Line, (SML), which describes the linear relationship between the security's return and its systematic risk, as measured by beta.

[2] Standard &Poor's, *The Stock Market Encyclopedia*, uses 5 years on monthly data to estimate beta coefficients.

The difficulty of measuring beta and its corresponding SML gave rise to extra-market measures of risk, found in the work of King (1966), Farrell (1974), Rosenberg (1974), Rosenberg and Marathe (1979), Rudd and Rosenberg (1979), and Rudd and Clasing (1982), Stone (1970, 1974), Ross (1976), Ross and Roll (1980), Blin and Bender in the APT Analytics Guide (2011) and Blin, Bender, and Guerard (1997) and culminated in the creation of the MSCI Barra and Sungard APT portfolio creation and management systems. We highlight the Barra Aegis system in this analysis. The Barra risk model was developed in the series of studies by Rosenberg and completely discussed in Rudd and Clasing (1982) and Grinhold and Kahn (2000). The extra-market risk measures are a seemingly endless source of discussion, debate, and often frustration among investment managers. Farrell (1974, 1997) estimated a four-"factor" extra-market model. Farrell took an initial universe of 100 stocks in 1974 (due to computer limitations), and ran market models for each stock to estimate betas and residuals from the market model:

$$R_{j_t} = a_j + b_j R_{\mathrm{M}_t} + e_j \tag{6.4}$$

$$e_{j_t} = R_{j_t} - \hat{a}_j - \hat{b}_j R_{\mathrm{M}_T}. \tag{6.5}$$

The residuals of (6.5) should be independent variables, if one factor (the market) is sufficient for modeling security returns. That is, after removing the market impact by estimating a beta, Farrell hypothesized that the residual of IBM should be independent of Dow, Merck, or Dominion Resources. The residuals should be independent, of course, with the market, in theory. The expected returns should be priced by only the beta. Farrell (1974) examined the correlations among the security residuals of (6.9) and found that the residuals of IBM and Merck were highly correlated, but the residuals of IBM and D (then Virginia Electric & Power) were not correlated. Farrell used a statistical technique known as Cluster Analysis to create clusters, or groups, of securities, having highly correlated market model residuals. Farrell found four clusters of securities based on his extra-market covariance. The clusters contained securities with highly correlated residuals that were uncorrelated with residuals of securities in the other clusters. Farrell referred to his clusters as "Growth Stocks" (electronics, office equipment, drug, hospital supply firms, and firms with above-average earnings growth), "Cyclical Stocks" (Metals, machinery, building supplies, general industrial firms, and other companies with above-average exposure to the business cycle), "Stable Stocks" (banks, utilities, retailers, and firms with below-average exposure to the business cycle), and "Energy Stocks" (coal, crude oil, and domestic and international oil firms).

Bernell Stone (1974) developed a two-factor index model which modeled equity returns as a function of an equity index and long-term debt returns. Both equity and debt returns had significant betas. In recent years, Stone and Guerard (2010a, b) have developed a portfolio algorithm to generate portfolios that have similar stock betas (systematic risk), market capitalizations, dividend yield, and sales growth cross sections, such that one can access the excess returns of the analysts' forecasts, forecast revisions, and breadth model, as one moves from low (least preferred) to high (most preferred) securities with regard to his or her portfolio construction variable (i.e., CTEF or a composite model of value and analysts' forecasting factors). In the Stone and Guerard (2010a) work, the ranking on forecasted return and grouping into fractile portfolios produce a set of portfolios ordered on the basis of predicted return score. This return cross section will almost certainly have a wide range of forecasted return values. However, each portfolio in the cross section will almost never have the same average values as that of the control variables. To produce a cross-sectional match on any of the control variables, we must reassign stocks. For instance, if we were trying to make each portfolio in the cross section that has the same average beta value, we could move a stock with an above-average beta value into a portfolio whose average beta value is below the population average. At the same time, we could shift a stock with a below-average beta value into the above-average portfolio from the below-average portfolio. The reassignment problem can be formulated as a mathematical assignment program (MAP). Using the MAP produces a cross-sectional match on beta or any other risk control variable. All (fractile) portfolios should have explanatory controls equal to their population average value.

In 1976, Ross published his "Arbitrage Theory of Capital Asset Pricing," which held that security returns were a function of several (4–5) economic factors. Ross and Roll (1980) empirically substantiated the need for 4–5 factors to describe the return generating process. In 1986, Chen, Ross, and Roll (CRR) developed an estimated multifactor security return model based on

$$R = a + b_{\mathrm{MP}}\,\mathrm{MP} + b_{\mathrm{DEI}}\,\mathrm{DEI} + b_{\mathrm{UI}}\,\mathrm{UI} + b_{\mathrm{UPR}}\,\mathrm{UPR} + b_{\mathrm{UTS}}\,\mathrm{UTS}\;te_t, \tag{6.6}$$

where MP is monthly growth rate of industrial production; DEI, change in expected inflation; UI, unexpected inflation; UPR, risk premium; and UTS, term structure of interest rates.

CRR defined unexpected inflation as the monthly (first) differences of the Consumer Price Index (CPI) less the expected inflation rate. The risk premia variable is the "Baa and under" bond return at time and less the long-term government bond return. The term structure variable is the long-term government bond return less the Treasury bill rates, known at time $t - 1$, and applied to time t. When CRR applied their five-factor model in conjunction with the value-weighted index betas, during the 1958–1984 period, the index betas are not statistically significant whereas the economic variables are statistically significant. The Stone, Farrell, and CRR multi-factor model used 4–5 factors to describe equity security risk. The models used different statistical approaches and economic models to control for risk.

The BARRA Model: The Primary Institutional Risk Model

As discussed previously, the most frequent approach for the prediction of risk is to use historical price behavior in the estimation of beta. Beta was defined as the sensitivity of the expected excess rate of return on the stock to the expected excess rate of return on the market portfolio. Unfortunately, the word *expected* has been used, and no good records of aggregate expectations exist. Thus, a major assumption has to be made to enable average (realized) rates of return to be used in place of expected rates of return, which, in turn, permits us to use the slope of regression line (estimated from realized data) to form the basis for a prediction of beta.

If this assumption, which essentially states that the future is going to be similar to the "average past," is made, then the estimation of historical beta proceeds as follows. Choose a suitable number of periods for which the excess returns of the security and market portfolio proxy are known. There is a subtle trade-off here. When more data points are used, the accuracy of the estimation procedure is improved, provided the relationship being estimated does not change. Usually the relationship does change; therefore, a small number of most recent data points is preferred so that dated information will not obscure the current relationship. It is usually accepted that a happy medium is achieved by using 60 monthly returns.[3] The security series is then regressed against the market portfolio series. This provides an estimate of beta (which is equivalent to the slope of the characteristic line) and the residual variance.

Menchero, Morozov, and Shepard (2010) used the CAPM framework and decompose the return of any asset into a systematic component, correlated with the market, and a residual uncorrelated with the market. The CAPM predicts that the residual return is zero. The predicted value of the residual does not preclude correlations among residual returns, because there may be multiple sources of equity return co-movement, even if there is a single source of expected return. It can be shown that if the regression equation is properly specified and certain other conditions are fulfilled, then the beta obtained is an optimal estimate (actually, minimum-variance, unbiased) of the true historical beta averaged over past periods. However, this does not imply that the historical beta is a good predictor of future beta. For instance, one defect is that random events impacting the firm in the past may have coincided with market movements purely by chance, causing the estimated value to differ from the true value. Thus, the beta obtained by this method is an estimate of the true historical beta obscured by measurement error. Rudd and Clasing (1982) discussed beta prediction with respect to the use of historic price information. Three possible prediction methods for beta were suggested. These are the following:

1. *Naïve*: $\hat{\beta}_j = 1.0$ for all securities (i.e., every security has the average beta).
2. *Historical*: $\hat{\beta}_j = \mathrm{H}\hat{\beta}_j$, the historical beta obtained as the coefficient forms an ordinary least squares regression.

[3] We have glossed over a number of econometric subtleties in these few sentences. Those readers who wish to learn more about these estimation difficulties are directed toward the following articles and the references contained there: Merton Miller and Myron Scholes, "Rates of Return in Relation to Risk: A Reexamination of Recent Findings," in *Studies in The Theory of Capital Markets*, ed. Michael Jensen (New York: Praeger Publishers, 1972), pp. 47–48.

3. *Bayesian-adjusted beta*: $\hat{\beta}_j = 1.0 + \text{BA}(\text{H}\hat{\beta}_j - 1)$, where the historical betas are adjusted toward the mean value of 1.0.

In each case, the prediction of residual risk is obtained by subtracting the systematic variance $\left(\hat{\beta}_j^2 V_{\text{M}}\right)$ from the total variance of the security. The residual variance is obtained directly from the regression.

However, relying simply upon historical price data is unduly restricting in that there are excellent sources of information which may help in improving the prediction of risk. For instance, most analysts would agree that fundamental information is useful in understanding a company's prospects. The *fundamental predictions of risk*, which were pioneered principally by Professor Barr Rosenberg and Vinay Marathe of the University of California at Berkeley, became the foundation of the Barra system.

The historical beta estimate will be an unbiased predictor of the future value of beta, provided that the expected change between the true value of beta averaged over the past periods and its value in the future is zero. If this expected change is not zero, then the historical beta estimate will be misleading (biased). Thus, if historical betas are used as a prediction of beta, there is an implicit assumption that the future will be similar to the past. Is this assumption reasonable? The answer is, probably not. The investment environment changes so rapidly that it would appear imprudent to use averages of historical (5-year) price data as predictions of the future.

Barr Rosenberg and Walt McKibben (1973) estimated the determinants of security betas and standard deviations. This estimation formed the basis of the Rosenberg extra-market component study (1974), in which security-specific risk could be modeled as a function of financial descriptors, or known financial characteristics of the firm. Rosenberg and McKibben found that the financial characteristics that were statistically associated with beta during the 1954–1970 period were:

1. Latest annual proportional change in earnings per share;
2. Liquidity, as measured by the quick ratio;
3. Leverage, as measured by the senior debt-to-total assets ratio;
4. Growth, as measured by the 5-year growth in earnings per share;
5. Book-to-Price ratio;
6. Historic beta;
7. Logarithm of stock price;
8. Standard deviation of earnings per share growth;
9. Gross plant per dollar of total assets;
10. Share turnover.

Rosenberg and McKibben used 32 variables and a 578-firm sample to estimate the determinants of betas and standard deviations. For betas, Rosenberg and McKibben found that the positive and statistically significant determinants of beta were the standard deviation of eps growth, share turnover, the price-to-book multiple, and the historic beta.[4] Rosenberg et al. (1975), Rosenberg and Marathe (1979), Rudd and Rosenberg (1979, 1980), and Rudd and Clasing (1982) expanded upon the initial Rosenberg MFM framework.

In 1975, Barr Rosenberg and his associates introduced the BARRA US Equity Model, often denoted USE1. We spend a great deal of time on the BARRA USE1 and USE3 models because 70 of the 100 largest investment managers use the

[4] When an analyst forms a judgment on the likely performance of a company, many sources of information can be synthesized. For instance, an indication of future risk can be found in the balance sheet and the income statement; an idea as to the growth of the company can be found from trends in variables measuring the company's position; the normal business risk of the company can be determined by the historical variability of the income statement; and so on. The approach that Rosenberg and Marathe take is conceptually similar to such an analysis since they attempt to include all sources of relevant information. This set of data includes historical, technical, and fundamental accounting data. The resulting information is then used to produce, by regression methods, the fundamental predictions of beta, specific risk, and the exposure to the common factors.

The fundamental prediction method of Barra starts by describing the company, see Rudd and Clasing (1982). The Barra USE1 Model estimated "descriptors," which are ratios that describe the fundamental condition of the company. These descriptors are grouped into six categories to indicate distinct sources of risk. In each case, the category is named so that a higher value is indicative of greater risk.

1. *Market variability.* This category is designed to capture the company as perceived by the market. If the market were completely efficient, then all information on the state of the company would be reflected in the stock price. Here the historical prices and other market variables are used in an attempt to reconstruct the state of the company. The descriptors include historical measures of beta and residual risk, nonlinear functions of them, and various liquidity measures.
2. *Earnings variability.* This category refers to the unpredictable variation in earnings over time, so descriptors such as the variability of earnings per share and the variability of cash flow are included.
3. *Low valuation and unsuccess.* How successful has the company been, and how is it valued by the market? If investors are optimistic about future prospects and the company has been successful in the past (measured by a low book-to-price ratio and growth in per share earnings), then the implication is that the firm is sound and that future risk is likely to be lower. Conversely, an unsuccessful and lowly valued company is more risky.
4. *Immaturity and smallness.* A small, young firm is likely to be more risky than a large, mature firm. This group of descriptors attempts to capture this difference.
5. *Growth orientation.* To the extent that a company attempts to provide returns to stockholders by an aggressive growth strategy requiring the initiation of new projects with uncertain cash flows rather than the more stable cash flows of existing operations, the company is likely to be more risky. Thus, the growth in total assets, payout and dividend policy, and earnings/price ratio is used to capture the growth characteristics of the company.
6. *Financial risk.* The more highly levered the financial structure, the greater is the risk to common stockholders. This risk is captured by measures of leverage and debt to total assets.

Finally industry in which the company operates is another important source of information. Certain industries, simply because of the nature of their business, are exposed to greater (or lesser) levels of risk (e.g., compare airlines versus gold stocks). Rosenberg and Marathe used indicator (dummy) variables for 39 industry groups as the method of introducing industry effects.

BARRA USE3 Model.[5] The BARRA USE1 Model predicted risk, which required the evaluation of the firm's response to economic events, which were measured by the company's fundamentals. Let us review the Barra prediction rules for the systematic risk and residual risk are expressed in terms of the descriptors, as discussed in Rudd and Clasing (1982). There are three major steps. First, for the time period during which the model is to be fitted, obtain common stock returns and company annual reports (for instance, from the COMPUSTAT database).[6] In order to make comparisons across firms meaningful, the descriptors must be normalized so that there is a common origin and unit of measurement, Table 6.1.

Table 6.1 Components of the risk indices

1. Index of market variability
Historical beta estimate
Historical sigma estimate
Share turnover, quarterly
Share turnover, 12 months
Share turnover, 5 years
Trading volume/variance
Common stock price (ln)
Historical alpha estimate
Cumulative range, 1 year
2. Index of earnings variability
Variance of earnings
Extraordinary items
Variance of cash flow
Earnings covariability
Earnings/price covariability
3. Index of low valuation and unsuccess
Growth in earnings/share
Recent earnings change
Relative strength
Indicator of small earnings/price ratio
Book/price ratio
Tax/earnings, 5 years
Dividend cuts, 5 years
Return on equity, 5 years
4. Index of immaturity and smallness
Total assets (log)
Market capitalization (log)
Market capitalization
Net plant/gross plant
Net plant/common equity

(continued)

[5] According to BARRA online advertisements.

[6] The COMPUSTAT database is one of the databases collected by Investors Management Sciences, Inc., a subsidiary of Standard & Poor's Corporation.

Table 6.1 (continued)

Inflation adjusted plant/equity
Trading recency
Indicator of earnings history
5. Index of growth orientation
Payout, last 5 years
Current yield
Yield, last 5 years
Indicator of zero yield
Growth in total assets
Capital structure change
Earnings/price ratio
Earnings/price, normalized
Typical earnings/price ratio, 5 years
6. Index of financial risk
Leverage at book
Leverage at market
Debt/assets
Uncovered fixed charges
Cash flow/current liabilities
Liquid assets/current liabilities
Potential dilution
Price-deflated earnings adjustment
Tax-adjusted monetary debt

Source: Rudd and Clasing, 1982, p. 114

The listing of the USE1 risk index components, as was reported in Rudd and Clasing (1982), was very informative. One wonders as to the weighting of the risk index components. The reader can find the variable weights in the risk index components in Rosenberg and Marathe (1976, see p 20). The Index of Market Variability was primarily determined by the historic Beta and the historic standard deviation of residual risk. The Index of Earnings Variability was primarily determined by the coefficient of variation of annual earnings per share in the last 5 years and the typical proportion of earnings that are extraordinary items. The Index of Unsuccess and Low Valuation was primarily determined by the measure of proportional change in adjusted earnings per share in the past two fiscal years and the "relative strength," the logarithmic rate of return, during the last year. The Index of Immaturity and Smallness was primarily determined by the ratio of gross plant to total equity and the logarithm of total assets. The Index of Growth Orientation was primarily determined by the normal value of the dividend yield during the last 5 years and the 5-year asset growth rate. The Index of Financial Risk was primarily determined by the total debt-to-assets ratio and the liquidity of the current financial position. The equations that formed the Index weights in USE1 were proprietary and undisclosed in USE2, USE3, and USE4.

In the Barra risk model, data is normalized. The normalization takes the following form. First, the "raw" descriptor Values for each company are computed. Next, the capitalization-weighted value of each descriptor for all the securities in the S&P 500 is computed and then subtracted from each raw descriptor. The transformed descriptors now have the property that the capitalization-weighted value for the S&P 500 stocks is zero. This step unambiguously fixes the "origin" for measurement; however, the unit of "length" is still arbitrary. To standardize the length, the standard deviation of each descriptor is calculated within a universe of large companies (defined as having a capitalization of $50 million or more). The descriptor is now further transformed by setting the value +1 to be one standard deviation above the S&P 500 mean (i.e., one unit of length corresponds to one standard deviation). Rudd and Clasing (1982) write

$$\mathrm{ND} = (\mathrm{RD} - \mathrm{RD}[\mathrm{S\&P}])/\mathrm{STDEV}[\mathrm{RD}], \tag{6.7}$$

where ND is the normalized descriptor value; RD the raw descriptor value as computer from the data; RD[S&P] the raw descriptor value for the (capitalization-weighted) S&P 500; and STDEV[RD] the standard deviation of the raw descriptor value calculated from the universe of large companies.

At this stage each company is indentified by a series of descriptors which indicate its fundamental position. If a descriptor value is zero, then the company is "typical" of the S&P 500 (for this characteristic) because the S&P 500 and the company both have the same raw value. Conversely, if the descriptor value is nonzero, then the company is atypical of the S&P 500, and this information may he used to adjust the prior prediction in order to obtain a better posterior prediction of risk.

In the second step, one groups the monthly data by quarters, and assemble the descriptors of each company as they would have appeared at the beginning of the quarter. The prediction rule is then fitted by linear regression which relates each monthly stock return in that quarter to the previously computed descriptors. These adjustments are combined as follows. Initially, in the absence of any fundamental information, the beta is set equal to its historical value. Then each descriptor is examined in turn, and if it is atypical, the corresponding adjustment to beta is made. For example, if two companies with the same historical beta are identical except that they have very different capitalizations, then one adjusts the risk of the large-capitalization company downward, relative to that of the small-capitalization company, because large companies typically have less risk than small companies. The fundamental knowledge of additional information improves the prediction of risk. The econometric prediction rule is similar; the prediction is obtained by adding the adjustments for all descriptors, in addition to the industry effect, to the historical beta estimate. The prediction rule for the beta of security i, in a given month, can be written as follows:

$$\hat{\beta}_i = \hat{b}_o + \hat{b}_I d_{Ii} + \ldots + \hat{b}_J d_{Ji}, \tag{6.8}$$

where $\hat{\beta}_i$ is the predicted beta; $\hat{b}_j$ the estimated response coefficients in the prediction rule; d_{ji} the normalized descriptor values for security i; and J the total number of descriptors.

In this prediction rule we can think of the first descriptor, d_{1i}, as the historical beta, $H\hat{\beta}$. Thus, if only the first descriptor is used, the prediction rule is similar to the specification of the Bayesian adjustment, (6.8). In this case, the linear regression provides estimates for $\hat{b}_o$ and $\hat{b}_1$, which indicate the optimal adjustment to historical beta for predictive purposes. Other descriptors in addition to historical beta are employed and appear in the prediction rule as d_{2i}. In other words, the fundamental predictions are direct generalizations of the "price only" predictions.

If the company is completely typical of market (i.e., the descriptors other than historical beta are all zero), then there is no further adjustment to the Bayesian-adjusted historical beta. This is intuitive; if the company is in no sense "special," then there is no reason to believe that the averaged true beta in the past will not equal the true beta in the future. However, if the company is atypical, then not all the descriptors (other than historical beta) will be zero. For simplicity, suppose that only the first (historical beta) and second descriptors are nonzero, where the latter has a value of one (i.e., this company is one standard deviation from the S&P 500 value). The prediction rule, (6.8), shows that the predicted beta is found by adding the adjustment $\hat{b}_2$ to the Bayesian-adjusted historical beta. In general, the total adjustment is the weighted sum of the coefficients in the prediction rule, where the weights are the normalized descriptor values which indicate the company's degree of deviance from the typical company.

In the third step, the Barra risk model estimates the company's exposure to each of the common factors and the prediction of the residual risk components. The first task is to form summary measures or indices of risk to describe all aspects of the company's investment risk. These are obtained by forming the weighted average of the descriptor values in each of the six categories introduced above, where the weights are the estimated coefficients from the prediction rule, (6.8), for systematic or residual risk. This provides six summary measures of risk, the risk indices, for each company. Again, these indices are normalized so that the S&P 500 has a value of zero on each index and a value of one corresponds to one standard deviation among all companies with capitalization of $50 million or more.

The prediction of residual risk is now found by performing a regression on the cross section of all security residual returns as the dependent variable where the independent variables are the risk indices.[7] The form of the regression, in a given month, is shown in (6.9):

$$r_i - \hat{\beta}_i r_M = c_1 \mathrm{RI}_{1i} + \ldots + c_6 \mathrm{RI}_{6i} + c_7 \mathrm{IND}_{1i} + \ldots + c_{45} \mathrm{IND}_{39,i} + u_i, \qquad (6.9)$$

where r_i is the excess return on security i; $\hat{\beta}_i$, the predicted beta, from (6.9); and R_M, the excess return on the market portfolio so that $r_i - \hat{\beta}_i r_M$ is the residual return on

[7] See Barr Rosenberg and Vinay Marathe, "Common Factors in Security Returns: Microeconomic Determinants and Macroeconomic Correlates," *Proceedings of the Seminar on the Analysis of Security Prices*, University of Chicago, May 1976, pp. 61–115 and Rosenberg and Marathe (1979).

security i; $RI_{1i}, \ldots, RI_{6i}$ are the six risk indices for security i, $IND_{1i}, \ldots, IND_{39,i}$ are the dummy variables for the 39 industry groups; u_i is the specific return for security i; and $c_1, \ldots, c_{45}$ are the 45 coefficients (factor returns) to be estimated.

The result from this cross-sectional regression is the specific return and specific risk on the security, together with the 45 coefficients. These estimated coefficients represent the returns that can be attributed to the factors in the month of the analysis.

The entire risk of the stock arises from two sources: the systematic or factor risk $\left(b_j^2 \text{ Var } [f]\right)$, and the nonfactor risk $\left(\sigma_j^2\right)$, the variance of the residual. In this case, however, the nonfactor risk is completely specific risk since no risk arises from interactions with other stocks. In other words, under these assumptions the single factor, f, is responsible for the only commonality among stock returns; thus, the random return component that is not related to the factor must be specific to the individual stock, j.

If we form a portfolio, P, with weights $h_{P1}, h_{P2}, \ldots, h_{PN}$, from N stocks, then the random excess return on the portfolio for a single factor is given by

$$R_P = \sum h_{Pj} r_j = \sum h_{Pj} b_j f + \sum h_{Pj} u_j = b_P f + \sum h_{Pj} u_j, \tag{6.10}$$

where $b_P = \sum h_{Pj} r_j$. The mean return and variance are

$$E[r_P] = a_P + b_P E[f],$$

where $a_P = \sum h_{Pj} a_j$, and

$$\text{Var}[r_P] = b_j^2 \text{Var}[f] + \sum h_P^2 \sigma_j^2, \tag{6.11}$$

where we have made use of the fact that the security-specific risk is *specific*, i.e., independent across stocks and independent of the factor return.

The market portfolio is just one particular portfolio. Let the security weights be $h_{M1}, h_{M2}, \ldots, h_{MN}$, and notice that $b_M = \sum h_{Mj} b_j$. We can set b_M to any value, and so we choose to set $b_M = 1$.[8] The market return statistics are then

$$E[r_M] = a_M + E[f]$$

and

$$\text{Var}[r_M] = \text{Var}[f] + \sum h_M^2 \sigma_j^2. \tag{6.12}$$

[8] This step is equivalent to defining an origin for measurement.

The regression coefficient of an individual stock's rate of return onto the market, or beta, is given by

$$\begin{aligned}\beta_j &= \mathrm{Cov}[r_j,\ r_\mathrm{M}]/\mathrm{Var}[r_\mathrm{M}]\\ &= \mathrm{Cov}\Big[b_jf + u_j,\ f + \sum h_{\mathrm{M}k}u_k\Big]/\mathrm{Var}[r_\mathrm{M}]\\ &= \left(b_j\mathrm{Var}[f] + h_{\mathrm{M}j}\sigma_j^2\right)/\mathrm{Var}[r_\mathrm{M}]\\ &= \left(b_j\mathrm{Var}[f] + h_{\mathrm{M}j}\sigma_j^2\right)\Big/\left(\mathrm{Var}[f] + \sum h_{\mathrm{M}_j}^2\sigma_j^2\right)\end{aligned} \tag{6.13}$$

so that

$$\beta_\mathrm{P} = \left(b_\mathrm{P}\mathrm{Var}[f] + \sum h_{\mathrm{M}j}h_{\mathrm{P}j}\sigma_j^2\right)\Big/\left(\mathrm{Var}[f] + \sum h_{\mathrm{M}_j}^2\sigma_j^2\right)^9.$$

Notice that the regression coefficient on the market and the regression coefficient on the factor (i.e., b_j and β_j, and b_P and β_P) are close but not identical. The difference lies in the last terms in the numerator and denominator in both cases. Where a single security is concerned, (6.13), the two sensitivities can only be equal when the market portfolio is composed of a single security; however, for a portfolio, the sensitivities will be close whenever the portfolio and market holdings are approximately equal (i.e., whenever $\sum h_{\mathrm{M}j}h_{\mathrm{P}j}$ is close to $\sum h_{\mathrm{M}j}^2$). In other words, for well-diversified portfolios (for instance, the majority of institutional portfolios) we may approximate the portfolio beta by its regression coefficient on the factor.

This approximation is useful for the analysis of residual return. Recall that the residual return of an individual portfolio (relative to the market portfolio) is equal to the total portfolio excess return on an equal-beta-levered market portfolio. That is, the residual return measures the return due to nonmarket strategy:

$$\text{Residual return} = r_\mathrm{P} - \beta_\mathrm{P}r_\mathrm{M}.$$

Thus, the residual variance is given by

$$\begin{aligned}w_\mathrm{P}^2 &= \mathrm{Var}[r_\mathrm{P} - \beta_\mathrm{P}r_\mathrm{M}]\\ &= \mathrm{Var}\Big[(b_\mathrm{P} - \beta_\mathrm{P})f + \sum (h_{\mathrm{P}j} - \beta_\mathrm{P}h_{\mathrm{M}j})u_j)\Big]\\ &= (b_\mathrm{P} - \beta_\mathrm{P})^2\mathrm{Var}[f] + \sum (h_{\mathrm{P}j} - \beta_\mathrm{P}h_{\mathrm{M}j})^2\sigma_j^2,\end{aligned} \tag{6.14}$$

since the nonfactor return, u_j, is uncorrelated with the factor return. Now, using the approximation that $\beta_\mathrm{P} = b_\mathrm{P}$, it follows that

$$w_\mathrm{P}^2 \cong \sum (h_{\mathrm{P}j} - \beta_\mathrm{P}h_{\mathrm{M}j})^2\sigma_j^2 = \sum \delta_{\mathrm{P}j}^2\sigma_j^2, \tag{6.15}$$

where $\delta_{Pj} = h_{Pj} - \beta_P h_{Mj}$. In other words, it is the discrepancy between portfolio and the holdings of the (equal-beta-levered) market portfolio that induces residual risk. In this formulation it is correct to write the sensitivity to the market as β_j since, by definition, a stock's beta is the exposure to the market. In addition, the nonmarket return is the expectation plus a random term with zero mean; i.e., nonmarket return is $\alpha_j + \varepsilon_j$, where $E[\varepsilon_j] = 0$, and α_j represents the expected abnormal rate of return, or alpha. That is, according to the stated assumptions of the single-factor model, the random nonmarket return on security j should be uncorrelated with the market return and similar returns on all other securities.

The mean excess return and variance for stock j are given by

$$E[r_j] = \sum_{k=1}^{K} b_{jk} E[f_k] + E[u_j]$$

and

$$\mathrm{Var}[rj] = \sum_{k=1}^{K} \sum_{l=1}^{K} b_{jk} b_{jl} \mathrm{Cov}[f_k, f_l] + \sigma_j^2, \tag{6.16}$$

where $\mathrm{Cov}[f_k, f_l]$ is the covariance between the factors and equals Var $[f_l]$ if $k = l$. This multiple factor model is specified by the security factor loadings, b_{jk}, and the factors, f_k.

If we now form a portfolio, P, with weights $h_{P1}, h_{P2}, \ldots, h_{PN}$, from N stocks, then the random excess return is given by

$$\begin{aligned} r_P &= \sum_{j=1}^{N} h_{Pj} r_j = \sum_{j=1}^{N} h_{Pj} \sum_{k=1}^{K} b_{jk} f_k + \sum_{j=1}^{N} h_{Pj} u_j \\ &= \sum_{k=1}^{K} \sum_{j=1}^{N} h_{Pj} b_{jk} f_k + \sum_{j=1}^{N} h_{Pj} u_j \\ &= \sum_{k=1}^{K} b_{Pk} f_k + \sum_{j=1}^{N} h_{Pj} u_j, \end{aligned} \tag{6.17}$$

where we have written $b_{Pk} = \sum h_{Pj} b_{jk}$ as the portfolio loading onto the kth factor. Since the market portfolio is a portfolio, the random excess return on the market is given by (6.16), with M replacing P; i.e.,

$$r_M = \sum_{k=1}^{K} b_{Mk} f_k + \sum_{j=1}^{N} h_{Mj} u_j.$$

Proceeding as before, the beta of the jth asset is given by

$$\begin{aligned}\beta_j &= \text{Cov}[r_j, r_m]/\text{Var}[r_M] \\ &= \left(\sum_{k=1}^{K}\sum_{l=1}^{K} b_{jk}b_{Ml}\text{Cov}[f_k, f_l] + b_{Mj}\sigma_j^2\right)/\text{Var}[r_M].\end{aligned} \tag{6.18}$$

It would appear that this complex expression is devoid of meaning; however, this is not the case. Consider the betas of the factors. In particular, for factor k

$$\begin{aligned}\beta_{fk} &= \text{Cov}[f_k, r_M]/\text{Var}[r_M] \\ &= \sum_{l=1}^{K} b_{Ml}\text{Cov}[f_k, f_l]/\text{Var}[r_M]\end{aligned}$$

and the beta of the specific component of return on the jth asset

$$\begin{aligned}\beta_{uj} &= \text{Cov}[u_j, r_M]/\text{Var}[r_M]. \\ &= h_M j\sigma_j^2/\text{Var}[r_M].\end{aligned}$$

That is, in the multiple factor model the security beta is a weighted average of the factor betas and the beta of the specific return of the security, where the weights are simply the factor loadings for the jth security. Notice that the beta of the stock's specific return is nonzero only because the security return is a component of the market return since the security is a part of the market. The intuition with which we wish to leave readers is that, far from being the primitive parameter in finance, the stock beta should be regarded as an average of a stock's exposures to a large number of factors influencing its return.

Now the residual return, the return due to a nonmarket strategy, on portfolio P is $r_P - \beta_P r_M$. Hence, the portfolio residual variance, w_P^2, is given by

$$\begin{aligned}w_P^2 &= \text{Var}[r_P - \beta_P r_M] \\ &= \text{Var}\left[\left\{\sum_{k=1}^{K}(b_{Pk} - \beta_P b_{Mk})f_k\right\} + \left\{\sum_{j=1}^{N}(h_{Pj} - \beta_P h_{Mj})u_j\right\}\right] \\ &= \text{Var}\left[\sum_{k=1}^{K}(\gamma_{Pk}f_k)\right] + \text{Var}\left[\sum_{j=1}^{N}\delta_{Pj}u_j\right],\end{aligned} \tag{6.19}$$

where γ is the Greek letter gamma and $\gamma_{Pk} = b_{Pk} - \beta_P b_{Mk}$ is the discrepancy in the portfolio factor loading and the equal-beta-levered market portfolio factor loading; δ_{Pj} is the discrepancy in the holdings, defined below (6.20), and the last step follows because the specific returns are uncorrelated with the factors.

Let the model for beta be given by

$$\beta_{nt} = b_0 + b_1 d_{1nt} + b_2 d_{2nt} + \ldots + b_J d_{Jnt} \tag{6.20}$$

for all time periods t and securities n, where the b's are coefficients for the systematic risk prediction rule and the d's are the J descriptor values for the nth company at time t. Further, let $E[\varepsilon_{nt}] = 0$ and $\mathrm{Cov}[\varepsilon_{nt}, r_{Mt}] = 0$ for all t, and define w_{nt}^2 to be the residual variance, i.e., $w_{nt}^2 = \mathrm{Var}[\varepsilon_{nt}]$. The model for residual risk is given by

$$w_{nt} = \bar{w}_t(s_0 + s_1 d_{1nt} + s_2 d_{2nt} + \ldots + s_J d_{Jnt}), \tag{6.21}$$

where $\bar{w}_t$ is the typical cross-sectional residual standard deviation in month t. This prediction rule is rewritten in terms of the mean absolute residual return, v_{nt}, for security n in month t and the typical mean absolute residual return in month t, $\bar{v}_t$. Therefore, $v_{nt} = E(|\varepsilon_{nt}|)$ and

$$v_{nt} = \bar{v}_t(s_0 + s_1 d_{1nt} + s_2 d_{2nt} + \ldots + s_J d_{Jnt}). \tag{6.22}$$

The estimate approach proceeds by substituting the beta prediction rule, (6.24), and then performing a "market conditional" regression for beta. The dependent variable is r_{nt}, and the independent variables are $d_{jnt} r_{Mt}$, so the model is

$$r_{nt} = \alpha + b_0(r_{Mt}) + b_1(d_{lnt} r_{MT}) + \ldots + b_J(d_{Jnt} r_{Mt}),$$

which provides preliminary estimates, $\hat{b}_0, \ldots, \hat{b}_j$. With these coefficients, the preliminary prediction of residual return is

$$\hat{\varepsilon}_{nt} = r_{nt} - (\hat{b}_0 + \hat{b}_1 d_{1nt} + \ldots + \hat{b}_J d_{Jnt}) r_{Mt}. \tag{6.23}$$

The next regression is fitted to estimate residual risk. It takes the form

$$|\hat{\varepsilon}_{nt}| = s_0(\hat{v}_t) + s_1(d_{1nt}\hat{v}_t) + \ldots + s_J(d_{Jnt}\hat{v}_t),$$

where

$$\bar{v}_t = \sum_{n=1}^{N} h_{Mnt} |\hat{\varepsilon}_{nt}|,$$

and h_{Mnt} is the proportion of security n in the market portfolio at time t. This regression provides estimates, $\hat{s}_0, \ldots, \hat{s}_J$.

The final step in this part of the analysis is to obtain prediction of systematic and residual risk by repeating these two regressions, but now using generalized least squares in order to correct for the different levels of residual risk across the securities.[9] The next task is to decompose the residual return into two components: specific return and the common factor return. This is achieved by a cross-sectional generalized least squares regression where the dependent variable is the residual return in month, t, $r_{nt} - \hat{\beta}_{nt} r_{Mt}$, and the independent variables are the risk indices and industry dummy variables. In this regression, each variable is weighted inversely to the predicted residual risk.

The statistically significant determinants of the security systematic risk became the basis of the BARRA E1 Model risk indexes. The domestic BARRA E3 (USE3, or sometimes denoted US-E3) model, with some 15 years of research and evolution, uses 13 sources of factor, or systematic, exposures. The sources of extra-market factor exposures are volatility, momentum, size, size nonlinearity, trading activity, growth, earnings yield, value, earnings variation, leverage, currency sensitivity, dividend yield, and non-estimation universe. The BARRA USE3 descriptors are included in the appendix to this chapter. We use the Barra USE3 Model to create portfolios using expected returns for equities in the United States for the 1980–2009 period.

Rudd and Clasing (1982) described the development and estimation of USE1. The MSCI Barra Model used in this chapter is the USE3 Model. The method of combining these descriptors into risk indices is proprietary to BARRA. There are 13 risk indexes or style factors in the USE3 Model. They are the following:

1. Volatility is composed of variables including the historic beta, the daily standard deviation, the logarithm of the stock price, the range of the stock return relative to the risk-free rate, the option pricing model standard deviation, and the serial dependence of market model residuals.
2. Momentum is composed of a cumulative 12-month relative strength variable and the historic alpha from the 60-month regression of the security excess return on the S&P 500 excess return.
3. Size is the log of the security market capitalization.
4. Size Nonlinearity is the cube of the log of the security market capitalization.
5. Trading Activity is composed of annualized share turnover of the past 5 years, 12 months, quarter, and month, and the ratio of share turnover to security residual variance.
6. Growth is composed of the growth in total assets, 5-year growth in earnings per share, recent earnings growth, dividend payout ratio, change in financial leverage, and analyst-predicted earnings growth.
7. Earnings Yield is composed of consensus analyst-predicted earnings to price and the historic earnings-to-price ratios.
8. Value is measured by the book-to-price ratio.

[9] This is the statistically efficient approach, and it requires that each observation be weighted inversely to its residual variance.

9. Earnings Variability is composed of the coefficient of variation in 5-year earnings, the variability of cash flow, and the variability of analysts' forecasts of earnings to price.
10. Leverage is composed of market and book value leverage, and the senior debt ranking.
11. Currency Sensitivity is composed of the relationship between the excess return on the stock and the excess return on the S&P 500 Index. These regression residual returns are regressed against the contemporaneous and lagged returns on a basket of foreign currencies.
12. Dividend Yield is the BARRA-predicted dividend yield.
13. Non-estimation Universe Indicator is a dummy variable which is set equal to zero if the company is in the BARRA estimation universe and equal to one if the company is outside the BARRA estimation universe.[10]

Stock Selection Modeling

This analysis builds upon Bloch, Guerard, Markowitz, Todd, and Xu (1993) and Guerard, Xu, and Gultekin (2012). We use the USER model described in Guerard et al. (2012). We refer the reader to these studies for much of the underlying expected returns literature. There are many approaches to security valuation and the creation of expected returns. The universe for all analysis consists of all securities on Wharton Research Data Services (WRDS) platform from which we download the CRSP database, I/B/E/S database, and the Compustat database. The I/B/E/S database contains consensus analysts' earnings per share forecast data and the Compustat database contains fundamental data, i.e., the earnings, book value, cash flow, depreciation, and sales data, used in this analysis for the December 1979–December 2007 time period. The stock selection model estimated in this study, denoted as the United States Expected Returns, USER, is

$$\begin{aligned}\mathrm{TR}_{t+1} = a_0 + a_1\mathrm{EP}_t + a_2\mathrm{BP}_t + a_3\mathrm{CP}_t + a_4\mathrm{SP}_t + a_5\mathrm{REP}_t + a_6\mathrm{RBP}_t \\ + a_7\mathrm{RCP}_t + a_8\mathrm{RSP}_t + a_9\mathrm{CTEF}_t + a_{10}\mathrm{PM}_t + e_t,\end{aligned} \tag{6.24}$$

where EP = [earnings per share]/[price per share] = earnings–price ratio; BP = [book value per share]/[price per share] = book–price ratio; CP = [cash flow per share]/[price per share] = cash flow–price ratio; SP = [net sales per share]/[price

[10] The Barra US Equity Model (USE4) was introduced in September 2011. The USE4 Model contains 12 style factors: Beta, Momentum, Size, Earnings Yield, Residual Volatility, Growth, Dividend Yield, Book-to-Price, Leverage, Liquidity, Nonlinear Size, and Nonlinear Beta. Menchero and Orr (2012) hold that the sample covariance matrix under-predicts risk and improved risk forecasts, lower biases, are linked to biases in eigenportfolios (removing eigenportfolio biases). Better risk-adjusted performance of portfolios results from better covariance adjustments.

per share] = sales–price ratio; REP = [current EP ratio]/[average EP ratio over the past 5 years]; RBP = [current BP ratio]/[average BP ratio over the past 5 years]; RCP = [current CP ratio]/[average CP ratio over the past 5 years]; RSP = [current SP ratio]/[average SP ratio over the past 5 years]; CTEF, consensus earnings-per-share I/B/E/S forecast, revisions and breadth; PM, Price Momentum; and e, randomly distributed error term.

The USER model is estimated cross-sectionally using a weighted latent root regression, WLRR, analysis on (6.24) to identify variables statistically significant at the 10% level; uses the normalized coefficients as weights; and averages the variable weights over the past 12 months, as described in Chapter 4.

The information coefficient, IC, is estimated as the slope of a regression line in which ranked subsequent returns are expressed as a function of the ranked strategy, at a particular point of time. In terms of information coefficients the use of the WLRR procedure produces the higher IC for the models during the 1998–2007 time period, 0.043, versus the equally weighted IC of 0.040, a result consistent with the previously noted studies. The IC test of statistical significance can be referred to as a Level I test. We have briefly surveyed the academic literature on anomalies and find substantial evidence that valuation, earnings expectations, and price momentum variables are significantly associated with security returns. Further evidence on the anomalies is found in Levy (1999)

Efficient Portfolio Construction Using the Barra Aegis System

The USER model can be input into the MSCI Barra Aegis system to create optimized portfolios. The equity factor returns f_k in the Barra United States Equity Risk Model, denoted USE3, are estimated by regressing the local excess returns r_n against the factor exposures, X_{nk},

$$r_n = \sum_{k=1}^{K_E} X_{nk} f_k + u_n. \tag{6.25}$$

The USE3 model uses monthly cross-sectional weighted regressions to estimate 13 (style) factors associated with extra-market covariances discussed earlier in the chapter. The USER model is our approximation of the expected return, or the forecast of active return, α, of the portfolio. Researchers in industry most often apply the Markowitz (1952) Mean-Variance framework to active management, as described in Grinold and Kahn (2000):

$$U = \alpha h - \lambda \omega^2 h^2. \tag{6.26}$$

Here α is the forecast of active return (relative to a benchmark which can be cash), ω is the active risk, and h is the active holding (the holding relative to the benchmark holding). The risk aversion parameter, λ, captures individual investor preference. By varying the tolerance or risk-aversion, λ, one can create the efficient Frontier in the Barra model. A similar procedure is used in Bloch et al. (1993). They created efficient portfolios by varying the pick parameter m which measured the risk-aversion. Grinold and Kahn (2000) use the Information Ratio, IR, as a portfolio construction objective to be maximized, which measures the ratio of residual return to residual risk:

$$\mathrm{IR} \equiv \frac{\alpha}{\omega}. \tag{6.27}$$

We construct an Efficient Frontier by varying the risk-aversion levels. The portfolio construction process uses 8% monthly turnover, after the initial portfolio is created, and 125 basis points of transaction costs each way. The USER-optimized portfolios outperform the market, defined here as the Russell 3000 Growth, R3G. The portfolio that maximizes the Geometric Mean (Markowitz 1976) and asset selection occurs at a risk-aversion level of 0.02. The Sharpe Ratio also is maximized at a risk acceptance parameter, RAP, of 0.02 with 109 stocks in the efficient portfolio.[11] A decreasing RAP implies that the more aggressive portfolios have a greater negative size exposure and implies that the portfolios contain smaller capitalized securities. A decreasing risk-aversion level produces a more concentrated portfolio, having fewer securities than a higher RAP portfolio, with the securities having smaller market capitalizations and higher exposures to momentum and growth. The efficient Frontier uses the Barra USE3 Short Model.

The efficient USER portfolio at a risk-aversion level of 0.02 offers exposure to MSCI Barra-estimated momentum, value, and growth exposures, see Table 6.2. The reader is hardly surprised with these exposures, given the academic literature and stock selection criteria and portfolio construction methodology employed.

The Guerard et al. (2012) USER analysis used the R3G benchmark, which began in December 1996. In this analysis, we can create a USER trade-off curve that covers the December 1979–December 2009 period by using the S&P as our benchmark. We find that the portfolio characteristics of the longer period analysis, 1980–2009, are very consistent with the portfolio characteristics of the 1997–2009 period, see Table 6.3. We find that an RAP of 0.001 is preferred for the 1980–2009 period.

The asset selection of the USER model is highly statistically significant and the risk index exposures are consistent with the shorter period.[12] The USER Efficient Frontier for the 1980–2009 period uses the Barra USE3L (United States Equity

---

[11] The regression-weighted USER outperforms the equally weighted model, EQ, in terms of maximizing the Sharpe Ratio, Information Ratio, Geometric Mean, and the t-value on Barra-estimated Asset Selection, a result consistent with Bloch et al. (1993), see Guerard et al. (2012).

[12] The statistical significance of USER in the 1980–2009 period is consistent with Bloch et al. (1993) and Stone and Guerard (2010b).

Table 6.2 USER efficient Frontier portfolio characteristics, 1996–2009

Benchmark: Russell 3000 Growth (R3G)

Transaction costs: 125 basis points each way

Risk acceptance parameter	0.015				0.020				0.050				0.100				0.200			
Portfolio	Mean	Return	IR	t	Mean	Return	IR	t	Mean	Return	IR	t	Mean	Return	IR	t	X	r	IR	t
Average number of assets	102				109				139				171				220			
Risk indices		2.11	0.31	1.12		1.96	0.31	1.11		1.53	0.30	1.09		1.31	0.33	1.19		1.03	0.33	1.21
Industries		−0.84	−0.24	−0.85		−0.77	−0.23	−0.82		−0.74	−0.28	−1.00		−0.20	−0.11	−0.39		0.10	0.04	0.16
Asset selection		2.26	0.46	1.68		2.61	0.53	1.92		2.52	0.58	2.08		2.06	0.55	1.97		2.51	0.73	2.65
Transaction cost		−2.61				−2.62				−2.59				−2.58				−2.58		
Total active		0.91	0.16	0.58		1.18	0.19	0.70		0.73	0.16	0.56		0.59	0.15	0.55		1.06	0.26	0.95
Total managed		4.09				4.36				3.91				3.77				4.24		
Currency sensitivity	0.04	0.01	0.03	0.11	0.03	0.02	0.10	0.35	0.02	−0.01	−0.01	−0.02	0.00	0.00	0.02	0.06	−0.01	0.01	0.05	0.19
Earnings variation	0.39	−0.33	−0.33	−1.18	0.36	−0.31	−0.33	−1.20	0.26	−0.16	−0.24	−0.86	0.19	−0.07	−0.13	−0.46	0.13	−0.01	−0.03	−0.10
Earnings yield	0.14	0.37	0.60	2.17	0.14	0.37	0.61	2.22	0.13	0.40	0.73	2.63	0.12	0.37	0.72	2.60	0.10	0.33	0.76	2.74
Growth	0.10	−0.14	−0.44	−1.61	0.10	−0.16	−0.52	−1.88	0.11	−0.16	−0.49	−1.78	0.10	−0.12	−0.36	−1.31	0.07	−0.07	−0.27	−0.97
Leverage	0.42	−0.03	−0.03	−0.12	0.40	−0.02	−0.03	−0.09	0.30	−0.02	−0.04	−0.14	0.22	0.01	−0.01	−0.02	0.15	0.01	−0.01	−0.04
Momentum	0.30	−0.58	−0.27	−0.96	0.29	−0.56	−0.27	−0.96	0.24	−0.52	−0.30	−1.08	0.20	−0.41	−0.28	−1.01	0.17	−0.30	−0.25	−0.90
Non-EST universe	0.44	−0.44	−0.14	−0.50	0.42	−0.42	−0.14	−0.50	0.32	−0.30	−0.13	−0.46	0.25	−0.22	−0.12	−0.43	0.18	−0.17	−0.12	−0.43
Size	−1.02	2.59	0.58	2.10	−0.94	2.42	0.58	2.11	−0.69	1.83	0.58	2.08	−0.52	1.34	0.55	2.00	−0.38	0.94	0.52	1.89
Size nonlinearity	−0.32	−0.02	−0.03	−0.10	−0.29	−0.01	−0.02	−0.08	−0.20	−0.02	−0.03	−0.12	−0.15	−0.02	−0.03	−0.12	−0.11	0.00	−0.01	−0.02
Trading activity	−0.64	0.70	0.25	0.91	−0.58	0.65	0.26	0.93	−0.42	0.47	0.25	0.92	−0.31	0.36	0.26	0.94	−0.22	0.26	0.26	0.94
Value	0.38	−0.18	−0.19	−0.67	0.36	−0.19	−0.21	−0.74	0.29	−0.12	−0.15	−0.55	0.25	−0.08	−0.13	−0.48	0.20	−0.08	−0.15	−0.53
Volatility	0.23	0.32	0.23	0.84	0.20	0.30	0.24	0.85	0.13	0.20	0.23	0.83	0.09	0.17	0.26	0.92	0.06	0.15	0.28	1.02
Yield	0.17	−0.15	−0.41	−1.49	0.16	−0.13	−0.39	−1.41	0.11	−0.05	−0.22	−0.79	0.09	−0.03	−0.15	−0.55	0.07	−0.03	−0.18	−0.65

Table 6.3 USER efficient Frontier portfolio characteristics, 1980–2009

Benchmark: S&P 500																				
Transaction costs: 125 basis points each way																				
Risk acceptance parameter		0.001				0.01				0.05				0.10				0.20		
	Information Mean	Return	Coefficient	*T*-statistic	Information Mean	Return	Coefficient	*T*-statistic	Information Mean	Return	Coefficient	*T*-statistic	Information Mean	Return	Coefficient	*T*-statistic	Information Mean	Return	Coefficient	*T*-statistic
Average number of assets	61.4				69.2				96.9				116.1				139.9			
Risk indices		1.45	0.18	0.96		1.59	0.21	1.14		1.10	0.19	1.03		0.99	0.20	1.09		0.83	0.20	1.12
Industries		−0.72	−0.17	−0.95		−0.55	−0.15	−0.80		−0.26	−0.08	−0.46		−0.02	−0.01	−0.04		0.13	0.06	0.31
Asset selection		5.24	0.84	4.62		4.31	0.77	4.22		3.50	0.76	4.17		2.70	0.65	3.54		2.62	0.71	3.89
Transaction cost		−2.78				−2.76				−2.72				−2.70				−2.69		
Total active		3.10	0.32	1.76		2.50	0.30	1.62		1.52	0.24	1.29		0.86	0.16	0.88		0.80	0.17	0.92
Total managed		14.27				13.67				12.69				12.04				11.97		
Barra risk indices																				
Currency sensitivity	0.06	0.03	0.08	0.46	0.06	0.02	0.07	0.36	0.05	0.01	0.04	0.19	0.05	0.01	0.03	0.15	0.04	0.01	0.05	0.27
Earnings variation	0.40	−0.33	−0.29	−1.57	0.31	−0.26	−0.30	−1.63	0.17	−0.12	−0.23	−1.26	0.11	−0.07	−0.18	−1.00	0.06	−0.04	−0.13	−0.70
Earnings yield	−0.03	−0.36	−0.60	−3.30	0.01	−0.04	−0.09	−0.48	0.03	0.09	0.20	1.10	0.03	0.07	0.20	1.10	0.03	0.06	0.20	1.10
Growth	0.43	−0.57	−0.49	−2.69	0.36	−0.44	−0.45	−2.48	0.26	−0.28	−0.41	−2.22	0.21	−0.22	−0.38	−2.10	0.16	−0.15	−0.34	−1.87
Leverage	0.47	−0.08	−0.07	−0.36	0.39	−0.05	−0.04	−0.22	0.26	−0.04	−0.04	−0.24	0.20	−0.03	−0.04	−0.19	0.14	−0.02	−0.04	−0.21
Momentum	0.53	−0.06	−0.01	−0.03	0.48	−0.04	0.00	0.00	0.38	−0.08	−0.03	−0.15	0.32	−0.04	−0.02	−0.08	0.27	−0.01	−0.01	−0.03
Non-EST universe	0.45	−0.13	−0.04	−0.22	0.42	−0.24	−0.07	−0.40	0.31	−0.17	−0.07	−0.39	0.25	−0.11	−0.05	−0.30	0.20	−0.05	−0.04	−0.20
Size	−1.63	4.02	0.57	3.15	−1.42	3.41	0.56	3.06	−0.94	2.11	0.50	2.74	−0.74	1.65	0.48	2.65	−0.57	1.30	0.48	2.63
Size non-linearity	−0.71	−0.58	−0.25	−1.39	−0.57	−0.36	−0.21	−1.16	−0.32	−0.14	−0.15	−0.80	−0.24	−0.09	−0.13	−0.69	−0.19	−0.09	−0.14	−0.76
Trading activity	−0.55	0.10	−0.03	−0.16	−0.56	0.11	−0.03	−0.15	−0.42	0.03	−0.05	−0.28	−0.35	0.05	−0.03	−0.16	−0.28	0.02	−0.04	−0.24
Value	0.15	−0.15	−0.16	−0.90	0.12	−0.15	−0.25	−1.36	0.06	−0.08	−0.21	−1.15	0.04	−0.08	−0.24	−1.33	0.03	−0.06	−0.22	−1.20
Volatility	0.58	−0.86	−0.12	−0.66	0.46	−0.70	−0.13	−0.72	0.26	−0.42	−0.14	−0.79	0.18	−0.31	−0.15	−0.80	0.13	−0.24	−0.16	−0.88
Yield	−0.39	0.42	0.35	1.94	−0.32	0.33	0.34	1.85	−0.20	0.20	0.31	1.68	−0.16	0.15	0.29	1.57	−0.14	0.10	0.23	1.28

Risk Model–Long) Risk Model. This chart shows the Frontier, reported in Miller, Xu, and Guerard (2012).

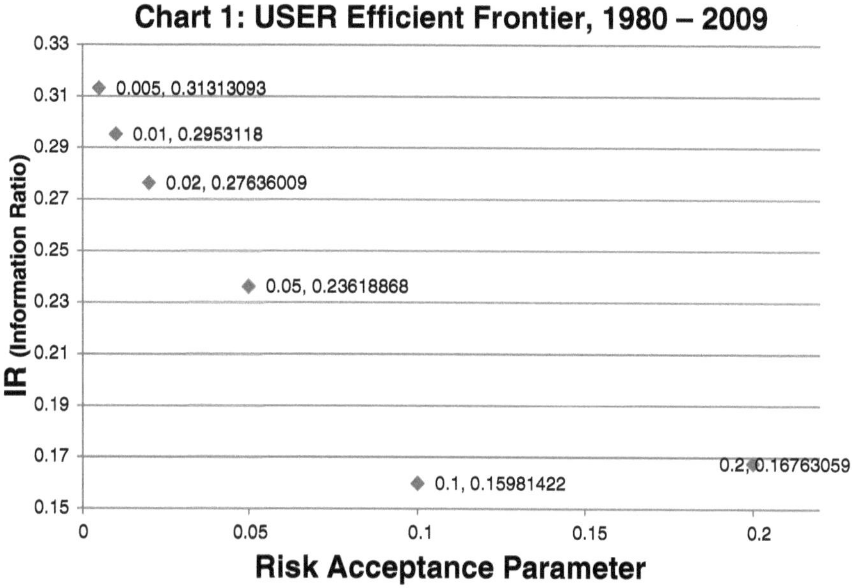

The creation of portfolios with a multifactor model and the generation of excess returns will hereby be referred to as a Level II test of portfolio construction.[13]

One could ask if the USER model resulted from a seemingly infinite number of variable tests and permutations. The USER was developed by the author in 1989 while at Drexel, Burnham, and Lambert in a consulting project for Continental Bank. Miller, Guerard, and Takano (1991) presented the initial model and the portfolio excess returns at the Berkeley Program in Finance meeting in Santa Barbara, in September 1990. Guerard worked for Harry Markowitz in the Global Portfolio Research Department, GPRD, at the Daiwa Securities Trust Company. The Continental Bank model was validated and expanded to test its use of 5-year relative variables and four-quarter variable weights lags. The Continental Bank model was validated in Bloch et al. (1993). Markowitz asked if the model could have been "in favor" or "unusually lucky" in its creation and initial implementation. Markowitz and Xu (1994)'s Data Mining Corrections (DMC) proposed three models to evaluate the outperformance of the best investment methodology when all of the back test data are available. It is human nature to be skeptical and wonder whether the best outperformance methodology is the result of "Data Mining." It has been applied routinely in the quantitative researches, for example, Bloch et al. (1993) and Guerard et al. (2010). This chapter follows previous papers doing the

[13] The eight-factor model generated statistically significant predictive power when used in the portfolio optimization and construction processes of Stone (1970, 1973, 2010a).

Table 6.4 US simulated returns: Jan 1980–Dec 2009

Portfolios	Monthly excess return to S&P 500 in percent	*t*-Statics
USER	0.28	1.72
BR1	0.16	1.29
BR2	0.13	1.12
RV1	0.22	1.48
RV2	0.04	0.32
FEP1	0.02	0.09
FEP2	−0.19	−0.87
CTEF	0.27	2.40
EP	0.09	0.50
BP	0.07	0.33
CP	0.16	0.90
SP	0.34	1.81
DP	0.22	1.21
PM71	0.16	0.84
PM	0.16	0.70
EWC	0.14	0.80
MQ	0.39	2.44

Data Mining Correction calculations with the longer data. We refer to the application of the Markowitz and Xu (1994) DMC test as a Level III test.

Fundamental factors like dividend-to-price (DP), earnings-to-price (EP) include forecast earnings-to-prices (FEP1, FEP2), book-to-price (BP), cash-to-price ratio (CP), sales-to-price ratio (SP), price momentums (PM71, PM) and financial analyst forecast earnings revisions (BR1, BR2, RV1, RV2) are not only used in risk modeling, e.g., Rosenberg (1974), but also used in stock selection models. Some researchers combine simple factors into a composite factor to enhance forecast power like USER and CTEF reported here. With the various expected return forecast model and risk model, researchers can pick a target portfolio from efficient Frontier according to preset investors' objectives. The excess returns of the portfolios created by the individual variables are denoted by model i. Here is the summary table, Table 6.4, of target portfolios generated by Barra Aegis optimization and portfolio management system, based on the previously discussed expected return "models," with the same risk trade-off parameter and the same trading cost.

The Markowitz and Xu (1994) DMC models assume that the *T* period backtest returns were identically and independently distributed (i.i.d.), and it is assumed that future returns are drawn from the same population (also i.i.d.). Let y_{it} be the logarithm of one plus the return for the *i*th portfolio selection methodology in period *t*. Then y_{it} is of the form

$$y_{ti} = \mu_i + \varepsilon_{it}, \tag{6.28}$$

where μ_i is a portfolio selection method effect and ε_{it} is a random deviation.

The random deviation ε_{it} has a zero mean and is uncorrelated with μ_i, i.e.,

$$E(\varepsilon_{it}) = 0 \tag{6.29}$$

$$\text{cov}(\mu_i,\ \varepsilon_{jt}) = 0 \text{ for all } i,\ j \text{ and } t. \tag{6.30}$$

The "best" linear unbiased estimate of the expected portfolio selection return vector μ is

$$\hat{\mu} = E(\mu)e + \text{Var}(\mu)\ \left[\tfrac{1}{T}C + \text{Var}(\mu)I\right]^{-1} \times (\tilde{y} - E(\mu)e), \tag{6.31}$$

where C is the covariance matrix of random effect, i.e.,

$$C = \text{cov}(\varepsilon_e,\ \varepsilon_j). \tag{6.32}$$

Markowitz and Xu (1994) refer to this as DMC Model III.
If one assumes that random effect is of form

$$\varepsilon_{it} = z_t + \eta_{it} \tag{6.33}$$

where Z_t is the period effect it is assumed to be uncorrelated with random effect η.
The best estimate of μ_i of (6.31) will be simplified to

$$\hat{\mu} = \bar{r} + \beta\ (\bar{r}_i - \bar{r}), \tag{6.34}$$

where

$$\bar{r} = \sum_{i=1}^{T} r_i/n. \tag{6.35}$$

That is the best estimate of means of return of portfolio selection i is not sample mean return, rather it is regressed back to the average return (the grand average). Markowitz and Xu (1994) refer to this as the DMC Model II and is the focus of their paper.

Model II can be used to test the null hypothesis that all these portfolios selected by different methods are equally good. If this hypothesis can be rejected, (6.35) gives the best estimate for each selected portfolio. In the above portfolios, the null hypothesis can be rejected with more than 90% confidence because the F-statistic equals 1.5 and β is estimated to be 0.33. Readers are referred to the original paper for detailed calculations.

DMC Model III Calculation

Instead of assuming that μ_i are random, Rao (1973) derived a formula for testing the significance of the null hypothesis that all means of these portfolios are the same. The F-statistic is calculated by

$$F = \frac{T-n+1}{n-1} \times \frac{T}{T-1} \times \left(\sum \sum c^{ij} \times \bar{r}_i \bar{r}_j - \frac{\left[\sum \sum c^{ij} (\bar{r}_i + \bar{r}_j) \right]^2}{4 \sum \sum c^{ij}} \right), \quad (6.36)$$

where (c^{ij}) is the inverse matrix of the C, the sample (estimated with T-1 D.F.) dispersion matrix as defined in (6.32).

When applying formula (6.36) to above portfolios, $F = 1.9$. Thus, we can reject the hypothesis with 95% confidence. The Bayesian estimate of means are the following:

Portfolio	$\bar{r}_i - \bar{r}$	Bayesian estimate of $\bar{r}_i - \bar{r}$	Estimate-to-actual ratio
S&P500	−0.09	−0.08	0.96
USER	0.14	0.12	0.86
BR1	0.06	0.05	0.84
BR2	0.03	0.02	0.59
RV1	0.07	0.09	1.18
RV2	−0.10	−0.08	0.82
FEP1	−0.15	−0.09	0.59
FEP2	−0.40	−0.32	0.79
CTEF	0.16	0.16	0.95
EP	−0.05	−0.05	1.05
BP	−0.10	−0.10	1.01
CP	0.02	0.02	0.82
SP	0.18	0.17	0.94
DP	0.09	0.09	0.96
PM71	−0.02	−0.03	1.26
PM71	−0.08	−0.09	1.16
EWC	0.00	−0.01	1.30
MQ	0.24	0.21	0.91

DMC provides some statistical answers to the impossible question whether an investment selection result is "lucky" or genuinely better. The DMC model III test produces a higher test statistic than DMC model II. The Bayesian's estimates are much closer to the simple sample estimates which ignore the other investment's influence. DMC model II is simpler and more plausible.

Conclusions

In this case study, we demonstrated the effectiveness of the Barra Aegis system to create investment management strategies to produce portfolios and attribute portfolio returns to the Barra multifactor risk model during the December 1979–December 2009 period. We find additional evidence to support the use of MSCI Bara multifactor models for portfolio construction and risk control. We report two results: (1) a composite model incorporating fundamental data, such as earnings, book value, cash flow, and sales, with analysts' earnings forecast revisions and price momentum variables to identify mispriced securities; (2) the returns to a multifactor risk-controlled portfolio allow us to reject the null hypothesis that results are due to data mining. We develop and estimate three levels of testing for stock selection and portfolio construction. The use of multifactor risk-controlled portfolio returns allows us to reject the null hypothesis that the results are due to data mining. The anomalies literature can be applied in real-world portfolio construction.

References

APT Analytics Guide 2011. SunGard APT, London www.sungard.com/apt/learnmore/

Beaton, A.E. and J. W. Tukey. 1974. "The Fitting of Power Series, Meaning Polynomials, Illustrated on Bank-Spectroscopic Data," *Technometrics* 16, 147–185.

Basu, S. 1977. "Investment Performance of Common Stocks in Relations to their Price Earnings Ratios: A Test of Market Efficiency." *Journal of Finance* 32, 663–682.

Bertsimas, D., G.J. Lauprete, and A. Samarov. 2004. "Shortfall as a Risk Measure: Properties, Optimzation and Applications". *Journal of Economic Dynamics and Control* 28, 1353–1381.

Bertsimas, D., C. Darnall, and R. Soucy. 1999. "Portfolio Construction Through Mixed-Integer Programming at Grantham, Mayo, Van Otterloo, and Company". *Interfaces* 29, 49–66.

Black, F., M.C. Jensen, and M. Scholes. 1972. "The Capital Asset Pricing Model: Some Empirical Tests," in *Studies in the Theory of Capital Markets,* edited by M. Jensen. New York: Praeger.

Blin, J.M. S. Bender, and J. B.Guerard, Jr. 1997. "Earnings Forecasts, Revisions and Momentum in the Estimation of Efficient Market-Neutral Japanese and U.S. Portfolios." In *Research in Finance* 15, edited by A. Chen.

Bloch, M., J.B. Guerard, Jr, H. M. Markowitz, P. Todd, and G-L. Xu. 1993. "A Comparison of Some Aspects of the U.S. and Japanese Equity Markets." *Japan &. the World Economy* 5, 3–26.

Brown, L.D. 1999 and 2008. Annotated I/B/E/S Bibliography.

Chan, L.K.C., Y. Hamao, and J. Lakonishok. 1991. "Fundamentals and Stock Returns in Japan". *Journal of Finance* 46, 1739–1764.

Chan, L.K.C., J. Lakonishok, and Y. Hamao 1993. "Can Fundamentals Predict Japanese Stock Returns" *Financial Analysts Journal.*

Chan, L.K.C., N. Jegadeesh and J. Lakonishok. 1996. "Momentum Strategies." *Journal of Finance* 51, 1681–1713.

Connor, G. 1993. "The Three Types of Factor Models: A Comparison of their Explanatory Power". Financial Analysts Journal 51, 42–46.

Connor, G. and R.A. Korajczyk. 1988. "Risk and Return in an Equilibrium APT: Application of a New Test Methodology." *Journal of Financial Economics* 21, 255–289.

Connor, G. and R.A. Korajczyk (1993). "A Test for the Number of Factors in an Approximate Factor Model." *Journal of Finance* 48, 1263–1291.

Connor, G. and R. A. Korajczyk . 1995. "The Arbitrage Pricing Theory and Multifactor Models of Asset Returns", Chapter 4 of *Finance, Handbooks in Operations Research and Management Science, Volume 9,* ed R. Jarrow, V. Maksimovic, and W. Ziemba. Amsterdam: North Holland, 1995

Connor, G. and R. A. Korajczyk. 2010. Factor Models in Portfolio and Asset Pricing Theory". In J. Guerard, Ed. The Handbook of Portfolio Construction: Contemporary Applications of Markowitz Techniques. New York: Springer.

Elton, E. J., M.J. Gruber, and M. Gultekin. 1981. "Expectations and Share Prices." *Management Science* 27, 975–987.

Elton, E.J., M.J. Gruber, S.J. Brown, and W.N. Goetzman. 2007. *Modern Portfolio Theory and Investment Analysis.* John Wiley &. Sons, Inc., Seventh Edition.

Fabozzi, F., F. Gupta &. Harry Markowitz . 2002. "The Legacy of Modern Portfolio Theory" *Journal of Investing,* 7–22.

Fama, E.F. and J.D. MacBeth. 1973. "Risk, Return, and Equilibrium: Empirical Tests." *Journal of Political Economy* 81, 607–636.

Fama, E.F. and K. R. French. 1992. "Cross-Sectional Variation in Expected Stock Returns." *Journal of Finance* 47, 427–465.

Fama, E.F. and K.R. French. 1995. "Size and the Book-to-Market Factors in Earnings and Returns. *Journal of Finance* 50, 131–155.

Fama, E.F., and K.R. French. 1996. "Multifactor Explanations of Asset Pricing Anomalies". *Journal of Finance* 51, 55–84.

Fama, E.F. and K. R. French. 2008. "Dissecting Anomalies". *Journal of Finance* 63, 1653–1678.

Farrell, J.L. Jr. 1974. "Analyzing Covariance of Returns to Determine Homogeneous Stock Groupings." *Journal of Business* 47, 186–207.

Farrell, J.L. Jr. 1997. *Portfolio Management: Theory and Applications.* New York McGraw-Hill/ Irwin.

Ferson, W.E. and C.R. Harvey. 1995. "Explaining the Predictability of Asset Returns." In *Research in Finance* 11, edited by A. Chen.

Ferson, W.E. "Theory and Empirical Testing of Asset Pricing Models." In R. Jarrow, V. Maksimovic, and W.T. Ziemba (eds.), Chapter 5, *Finance, Handbooks in Operations Research and Management Science,* 9, Amsterdam: North Holland, 1995.

Ferson, W.E. and C.R. Harvey. 1998. "Fundamental Determinants of National Equity Market Returns: A Perspective on Conditional Asset Pricing." *Journal of Banking &. Finance* 21, 1625–1665.

Fisher, L. and J. H. Lorie. 1970. "Some Studies of Variability of Returns on Investments in Common Stocks." *Journal of Business* 43, 39–143.

Fisher, L. and J. H. Lorie. 1964. "Rates of Return on Investments in Common Stocks." *Journal of Business 7,* 1–21.

Grinold, R and R. Kahn. 1999. *Active Portfolio Management* New York: McGraw-Hill/Irwin.

Guerard, J.B., Jr., M. Gultekin, and B.K. Stone. 1997. "The Role of Fundamental Data and Analysts' Earnings Breadth, Forecasts, and Revisions in the Creation of Efficient Portfolios." In *Research in Finance* 15, edited by A. Chen.

Guerard, J.B. Jr., M Takano, and Y. Yamane. 1993. "The Development of Efficient Portfolios in Japan with Particular Emphasis on Sales and Earnings Forecasting." *Annals of Operations Research* 45, 91–108.

Guerard, J.B. 1997. "Is There a Cost to be Socially Responsible in Investing?" *Journal of Forecasting* 16, 475–490.

Guerard, J.B., Jr. 2006. "Quantitative Stock Selection in Japan and the US: Some Past and Current Issues." *Journal of Investing,* 15.

Guerard, J.B., S. C. Chettiappan, and G-L, Xu. 2010. "Stock-Selection Modeling and Data Mining Corrections: Long-Only versus 130/30 Models". *The Handbook of Portfolio Construction: Contemporary Applications of Markowitz Techniques.* New York: Springer.

Guerard, J.B. 2010. *The Handbook of Portfolio Construction: Contemporary Applications of Markowitz Techniques.* New York: Springer. Chapter 3.

Guerard, J.B., Jr., M.N. Gultekin, and G. Xu. 2012. "Investing with Momentum: The Past, Present, and Future", *Journal of Investing* 21, 68–80.

Guerard, J.B. 2012. Global Earnings Forecast Efficiency (2012), *Research in Finance* 28, 19–47.

Gültekin, M. N. and N. B. Gültekin. 1987. "Stock Return Anomalies and Tests of the APT." *Journal of Finance* 42, 1213–1224.

King, B.F. 1966. "Market and Industry Factors in Stock Price Behavior." *Journal of Business* 39, 139–191.

Latane, H.A. 1959."Criteria for Choice Among Risky Ventures". *Journal of Political Economy* 67, 144–155.

Levy, H. 1999. Introduction to Investments. Cincinnati: South-Western College Publishing. Second Edition.

Lintner, J. 1965. "The Valuation of Risk Assets on the Selection of Risky Investments in Stock Portfolios and Capital Investments." *The Review of Economics and Statistics,* 13–37.

Markowitz, H.M. 1952. "Portfolio Selection." *Journal of Finance* 7, 77–91.

Markowitz, H. M. 1956. The Optimization of a Quadratic Function Subject to Linear Constraints. *Naval Research Logistics Quarterly* 3, 111–133.

Markowitz, H.M. 1959. *Portfolio Selection: Efficient Diversification of Investment.* Cowles Foundation Monograph No. 16. New York, John Wiley &. Sons.

Markowitz, H.M. 1976. "Investment in the Long Run: New Evidence for an Old Rule," *Journal of Finance* 31, 1273–1286.

Markowitz, H.M. *Mean-Variance Analysis in Portfolio Choice and Capital Markets.* Oxford: Basil Blackwell, 1987.

Markowitz, H.M., and G. Xu. 1994. "Data Mining Corrections." *Journal of Portfolio Management* 21, 60–69.

Markowitz, H.M. 2000. *Mean-Variance Analysis in Portfolio Choice and Capital Markets.* New Hope, PA: Frank J. Fabozzi Associates.

Menchero, J., A. Morozov, and P.Shepard. 2010. "Global Equity Risk Modeling", in J.B. Guerard, Ed. *The Handbook of Portfolio Construction: Contemporary Applications of Markowitz Techniques.* New York: Springer.

Menchero, J. and D.J. Orr. 2012. "Improving Risk Forecasts for Optimized Portfolios". *Financial Analysts Journal* 68, 40–50.

Miller, J. D., J.B. Guerard, Jr. and M. Takano. 1991. "Bridging the Gap Between Theory and Practice in Equity Selection Modeling: Case Studies of U.S. and Japanese Models. An earlier version was presented at the Berkeley Program in Finance. September 1990. Santa Barbara, CA.

Miller, D., G. Xu, and Guerard, J.B. 2012. "Portfolio Construction and Management in the Barra Aegis System: A Case Study using the USER Data", Working Paper.

Mossin, J. 1966. "Equilibrium in a Capital Asset Market." *Econometrica* 34, 768–783.

Pastor, L. and R. F. Stambaugh. 2003. "Liquidity Risk and Expected Stock Returns." *Journal of Political Economy,* 111, 642–684.

Rao, C.R. 1972. "Estimation of Variance and Covariance Components in Linear Models". *Journal of the American Statistical Association* 67, 112–115.

Rao, CR. *Linear statistical Inference and Its Applications.* New York: John Wiley & Sons, 1973, 2nd edition

Rachev, S. and S. Mittnik. 2000. *Stable Paretian Models in Finance.* New York: Wiley.

Rudd, A. and B. Rosenberg. 1980. "The 'Market Model' in Investment Management." *Journal of Finance* 5, 597–607.

Rosenberg, B., M. Hoaglet, V. Marathe and W. McKibben. 1975. "Components of Covariance in Security Returns." Working paper no. 13, Research Program in Finance. Institute of Business and Economic Research. University of California. Berkeley.

Rosenberg, B. and V. Marathe, 1979. "Tests of Capital Asset Pricing Hypotheses." In *Research in Finance* 1,1979, edited by H. Levy.

Rudd, A, and B. Rosenberg. 1979. "Realistic Portfolio Optimization." In *Portfolio Theory, 25 Years After.* Edited by E. Elton and M. Gruber. Amsterdam: North-Holland.

Ramnath, S., S. Rock, and P. Shane. 2008. "The Financial Forecasting Literature: A Taxonomy with Suggestions for Further Research." *International Journal of Forecasting,* 24, 34–75.

Rosenberg, B. and W. McKibben. 1973. "The Prediction of Systematic and Specific Risk in Common Stocks." *Journal of Financial and Quantitative Analysis* 8, 317–333.
Rosenberg, B. 1974. "Extra-Market Components of Covariance in Security Returns." *Journal of Financial and Quantitative Analysis* 9, 263-214.
Rosenberg, B. and V. Marathe. 1975. "Tests of Capital Asset Pricing Hypotheses". Working Paper, No. 32. Research Program in Finance. Institute of Business and Economic Research. Berkeley, University of California.
Rosenberg, B. and V. Marathe. 1976. "Common Factors in Security Returns: Microeconomic Determinants and Macroeconomic Correlates". Working Paper, No. 44. Research Program in Finance. Institute of Business and Economic Research. Berkeley, University of California.
Rosenberg, B. and V. Marathe, 1979. "Tests of Capital Asset Pricing Hypotheses." In *Research in Finance* 1, 1979, edited by H. Levy.
Ross, S.A. 1976. "The Arbitrage Theory of Capital Asset Pricing." *Journal of Economic Theory* 13, 341–360.
Roll, R. 1969. "Bias in Fitting the Sharpe Model to Time series Data", *Journal of Financial and Quantitative Analysis* 4, 271–289.
Roll, R. 1977. "A Critique of the Asset Pricing Theory's Tests". *Journal of Financial Economics* 4, 129–176.
Ross, S.A. and R. Roll. 1980. "An Empirical Investigation of the Arbitrage Pricing Theory." *Journal of Finance* 35, 1071–1103.
Rudd, A., and H. K.Clasing 1982. *Modern Portfolio Theory: The Principles of Investment Management.* Homewood, IL: Dow-Jones Irwin.
Shanken, J. 1985. "Multi-Beta CAPM or Equilibrium APT: A Reply." *Journal of Finance* 40, 1189–1196.
Shanken, J. 1992. "The Current State of the Arbitrage Pricing Theory." *Journal of Finance* 47, 1569–1574.
Sharpe, W.F. 1963. "A Simplified Model for Portfolio Analysis." *Management Science* 9, 277–293.
Sharpe, W.F. 1964. "Capital Asset Prices: A Theory of Market Equilibrium under Conditions of Risk." *Journal of Finance* 19, 425–442.
Sharpe, W.F. 1971. "Mean-Absolute-Deviation Characteristic Lines for Securities and Portfolios", Management Science 18, B1-B13.
Solnik, B. 1974. "Why Not Diversify Internationally Rather than Domestically". *Financial Analysts Journal* 38, 48–54.
Solnik, B. 2000. *International Investments,* Reading, Mass: Addison Wesley Longman, Inc., 4th Edition.
Stone, B.K. 1970. *Risk, Return, and Equilibrium: A General Single-Period Theory of Asset Selection and Capital Market Equilibrium.* Cambridge, MA: MIT Press.
Stone, B. K. 1973. "A Linear Programming Formulation of the General Portfolio Selection Problem." *Journal of Financial and Quantitative Analysis,* 8, 621–636.
Stone, B.K. 1974. "Systematic Interest-Rate Risk in a Two-Index Model of Returns." *Journal of Financial and Quantitative Analysis* 9, 709–721.
Stone, B.K., J.B. Guerard, Jr., M. Gultekin, and G. Adams. 2002. "Socially Responsible Investment Screening." Working Paper. Marriott School of Management. Brigham Young University.
Stone, B. K. 2003. "Rank-Based Security Grouping, Multicolinearity, and the Assessment of Cross-Sectional Return Dependencies," Working Paper. The Marriott School of Management, Brigham Young University.
Stone, B. K. and J.B. Guerard, Jr. 2010. Methodologies for Isolating and Assessing the Portfolio Performance Potential of Stock Return Forecast Models with an Illustration, in J.B. Guerard, Jr., editor, *The Handbook of Portfolio Construction: Contemporary Applications of Markowitz Techniques.* New York: Springer.
Stone, B. K. and J.B. Guerard, Jr. 2010b. "Quantitative Stock Selection in Japan and the United States: Some Past and Current Issues." Proceedings, 50th Annual CRSP Forum (Chicago, November 2010)

Chapter 7
More Markowitz Efficient Portfolios Featuring the USER Data and an Extension to Global Data and Investment Universes

In the previous chapter, we used the Barra Aegis system to create and measure portfolios using the USER model. The Barra Model is referred as a fundamental risk model because security fundamental data is used to create the risk, or style, indexes. In this chapter, we create portfolios using statistically-based risk models in the USA and global markets. In this chapter, we address several additional issues in portfolio construction and management with Guerard, Xu, and Gultekin (2012) USER data. First, we test the issue of whether Markowitz mean–variance, MV, portfolio construction model (1956, 1959, 1987), with a fixed upper bound on security weights, dominates the Markowitz enhanced index tracking, EIT, portfolio construction model (1987) in which security weights are an absolute deviation from the security weight in the index. We will refer to the absolute deviation from the benchmark weight-enhanced index portfolio construction weight as the equal active weighting, or EAW, portfolio construction model. Guerard, Krauklis, and Kumar (2012) reported that MV portfolios produced higher Information Ratios and Sharpe Ratios than EAW portfolios with weights less than EAW4. A newer approach to the systematic risk optimization technique is the Systematic Tracking Error optimization technique reported by Wormald and van der Merwe (2012). We will show the effectiveness of the Systematic Tracking Error approach using Global Expected Returns (GLER) data over the 2002–2011 period. Finally, we demonstrate using the Axioma system and its Alpha Alignment Factor (AAF) analysis reported in Saxena and Stubbs (2012) that the AAF is appropriate for USER and GLER Data and that the Axioma Statistical Risk Model dominates the Axioma Fundamental Model.[1]

---

[1] In Chap. 6, we reported that asset selection was statistically significant in the Barra Aegis system. We report similar results with Sungard APT and Axioma. The author's belief is that the three systems can be used to produce highly statistically significant asset selection and very good portfolio returns and great risk-return statistics. One needs to decide if one wants to set Lambda, as with Sungard APT, active risk, as with Axioma, and risk acceptance parameters, as with Barra. In the author's view, APT, the system that the author has used since 1989 is outstanding and very adequate. Many (intelligent) people choose active risk (tracking error targets). As long as you are statistically significant in asset selection with the USER variable (or other proprietary forms) and

J.B. Guerard, Jr., *Introduction to Financial Forecasting in Investment Analysis*, DOI 10.1007/978-1-4614-5239-3_7,

The security weights are the primary decision variables to be solved in efficient portfolios. Second, we test whether a (traditional) mean–variance optimization technique using the portfolio variance as the relevant risk measure dominates risk-return trade-off curve using the Blin-Bender APT Tracking Error at Risk (TaR) optimization technique which emphasizes systematic, or market, risk. The APT measure of portfolio risk, TaR, estimates the magnitude that the portfolio return may deviate from the benchmark return over 1 year. Specifically, the TaR optimization technique emphasizes systematic risk, rather than total risk, in portfolio optimization. A statistically-based principal components analysis (PCA) model is used to estimate and monitor portfolio risk in the Blin and Bender TaR system.

To address these issues, we construct efficient portfolios with the USER data, solving for security weights using mean–variance and equal active weighting portfolio construction models for the 1997–2009 period. The MV portfolio construction model with fixed security upper bounds performs very well in comparison to EAW portfolio construction models. Mean–variance portfolios with a 4% security upper bound outperform EAW 1, 2, and 3% strategies. One must use an (at least) EAW 4% strategy to outperform the MV portfolio construction model with a 4%, see Guerard, Krauklis, and Kumar (2012). Index-tracking portfolio construction models are extremely useful if a manager is more concerned with underperforming an index; however, the portfolio manager must be aggressive with the EAW strategy to outperform a traditional mean–variance portfolio construction analysis.

We employ mean–variance and TaR optimization techniques to test whether equal active weighting strategies of 1, 2, 3, 4, and 5% (weight deviations from the index, or benchmark, weights) outperform mean–variance strategies using 4 and 7% maximum security weights. We will show mean–variance portfolios using the Tracking Error at Risk optimization technique outperform the mean–variance optimization technique during the 1997–2009 period. Both optimization techniques produce statistically significant asset selection. We employ the Wormald and van der Merwe (2012) Systematic Tracking Error optimization techniques and find statistically significant asset selection. In this chapter, we examine two portfolio construction models: mean–variance and equal active weighting models; and two portfolio optimization techniques: mean–variance and Tracking Error at Risk, and Systematic Tracking Error optimization techniques.

Lambda is a measure of the trade-off between expected returns and risk, as measured by the portfolio standard deviation. Generally, the higher the lambda, the higher is the expected ratio of expected return to standard deviation. That is,

are man-enough to implement the model to maximize the Sharpe Ratio and Geometric Mean (having a negative size exposure and positive momentum, growth, and value exposures), then the choice of APT and Axioma (and Barra) is analogous to the man who is asked if he prefers blondes, brunettes, or redheads; one prefers great minds, strong wills, good looks, and the hair color, preferably natural, is a lesser concern. Not all risk models and optimizers work, as we found out in the McKinley Capital Horse Race and research seminars of 2009 and 2011. Some systems are more expensive and their portfolios are dominated by APT, Axioma, and Barra on a risk-return analysis. We found a decidedly negative correlation between cost and performance.

creating portfolios with less than optimal lambdas produce portfolio excess returns that are not statistically different from zero, whereas appropriate lambdas create portfolios that are statistically significant. In the King's English, benchmark-hugging portfolio construction techniques can destroy significant asset selection. We assume that the portfolio manager seeks to maximize the combination of portfolio Geometric Mean (GM), Sharpe Ratio (ShR), and Information Ratio (IR), and asset selection in the Barra attribution analysis. If a portfolio manager has models that produce slightly different ordering on these criteria, we maximize the Geometric Mean (Latane 1959, Latane, Tuttle, and Jones 1975, Markowitz (1959, 1976), and Vander Weide 2010) as the ultimate criteria, since it is well known that risk is implicit in the Geometric Mean (Markowitz, Chap. 9).

Constructing Efficient Portfolios

In the previous chapter, we discussed the Barra Aegis system and its use in creating efficient portfolios that produce statistically significant asset selection. Let us step back for a moment and review six decades of portfolio construction and management. In the beginning, there was Markowitz (1952). The Markowitz portfolio construction approach seeks to identify the efficient frontier, the point at which returns are maximized for a given level of risk, or minimize risk for a given level of return. The reader is referred to Markowitz (1959) for the seminal discussion of portfolio construction and management. The portfolio expected return, $E(R_{\mathrm{p}})$, is calculated by taking the sum of the security weights, w, multiplied by their respective expected returns. The portfolio standard deviation is the sum of weighted security covariances.

$$E(R_{\mathrm{p}}) = \sum_{i=1}^{N} w_i E(R_i), \tag{7.1}$$

$$\sigma_{\mathrm{p}}^2 = \sum_{i=1}^{N} \sum_{j=1}^{N} w_i w_j \sigma_{ij}, \tag{7.2}$$

where $\sum_{i=1}^{N} w_i = 1$ the security weighting summing to one indicates that the portfolios are fully invested.

The Markowitz framework measured risk as the portfolio standard deviation, its measure of dispersion, or total risk. One seeks to minimize risk, as measured by the covariance matrix in the Markowitz framework, holding constant expected returns. Elton, Gruber, Brown, and Goetzmann (2007) write a more modern version of the traditional Markowitz mean–variance problem as a maximization problem:

$$\theta = \frac{E(R_{\mathrm{p}}) - R_{\mathrm{F}}}{\sigma_{\mathrm{p}}^2}, \tag{7.3}$$

where $\sum_{i=1}^{N} w_i = 1$
and

$$\sigma_{\mathrm{p}}^2 = \sum_{i=1}^{N} w_i^2 \sigma_i^2 + \sum_{i=1}^{N} \sum_{j=1}^{N} w_i w_j \sigma_{ij}, \quad i \neq j$$

and R_{F} is the risk-free rate (90-day treasury bill yield).

The optimal portfolio weights are given by:

$$\frac{\partial \theta}{\partial w_i} = 0.$$

As in the initial Markowitz analysis, one minimizes risk by setting the partial derivative of the portfolio risk with respect to the security weights, the portfolio decision variables, to 0.

Modern portfolio theory evolved with the introduction of the Capital Asset Pricing Model, the CAPM. Implicit in the development of the CAPM by Sharpe (1964), Lintner (1965), and Mossin (1966) is that the investors are compensated for bearing systematic or market risk, not total risk. Systematic risk is measured by the stock beta. The beta is the slope of the market model in which the stock return is regressed as a function of the market return.[2] An investor is not compensated for bearing risk that may be diversified away from the portfolio.

The CAPM holds that the return to a security is a function of the security's beta.

$$R_{jt} = R_{\mathrm{F}} + \beta_j [E(R_{Mt}) - R_{\mathrm{F}}] + e_j, \tag{7.5}$$

where R_{jt} = expected security return at time t; $E(R_{Mt})$ = expected return on the market at time t; R_{F} = risk-free rate; β_j = security beta; and e_j = randomly distributed error term.

An examination of the CAPM beta, its measure of systematic risk, from the Capital Market Line equilibrium condition follows.

$$\beta_j = \frac{\mathrm{Cov}(R_j, R_M)}{\mathrm{Var}(R_M)}. \tag{7.6}$$

The difficulty of measuring beta and its corresponding SML gave rise to extra-market measures of risk found in the work of Rosenberg (1974), Rosenberg and Marathe (1979), Ross (1976), and Ross and Roll (1980).[3] The fundamentally-based domestic Barra risk model was developed in the series of studies by Rosenberg and

[2] Harry Markowitz often (always) reminds his audiences and readers that he discussed the possibility of looking at security returns relative to index returns in Chap. 4, footnote 1, page 100, of Portfolio Selection (1959).

[3] The reader is referred to Chap. 2 of Guerard (2010) for a history of multi-index and multi-factor risk models.

thoroughly discussed in Rudd and Clasing (1982) and Grinhold and Kahn (1999), and as discussed in the previous chapter.

The total excess return for a multiple-factor model (MFM) in the Rosenberg methodology for security j, at time t, dropping the subscript t for time, may be written like this:

$$E(R_j) = \sum_{k=1}^{K} \beta_{jk} \tilde{f}_k + \tilde{e}_j. \tag{7.7}$$

The nonfactor, or asset-specific return on security j, is the residual risk of the security after removing the estimated impacts of the K factors. The term f is the rate of return on factor "k." A single-factor model, in which the market return is the only estimated factor, is obviously the basis of the CAPM. Accurate characterization of portfolio risk requires an accurate estimate of the covariance matrix of security returns. An alternative to the fundamentally-based Barra risk model is a risk model based on statistically-estimated (orthogonal) principal components, as described in the APT model of Blin, Bender, and Guerard (1997).

Extensions to the Traditional Mean–Variance Model

A second extension to the mean–variance approach involves the minimization of the tracking error of an index. Markowitz (1987, 2000) rewrites the general portfolio construction model variance, V, to be minimized as:

$$V = (X - W)^T C(X - W), \tag{7.8}$$

where $W^T = (W_1, \ldots, W_n)$ = the weights of an index of returns, X are the portfolio weights, and $r^T = (r_1, \ldots, r_n)$ = security returns.

One creates portfolios by allowing portfolio weights to differ from index weights by ±1%, EAW1, 2%, EAW2, 3%, EAW3, 4%, EAW4, or 5%, EAW5. Obviously, one can use an infinite set of EAW variations. We restrict this analysis to EAW1, EAW2, EAW3, and EAW4 for simplicity.

Portfolio Construction, Management, and Analysis: An Introduction to Tracking Error at Risk

The USER simulation conditions are identical to those described in Guerard et al. (2012), in which we use monthly optimization with 8% turnover, 125 basis points, each way, of transactions cost.[4] We use the APT risk model and optimizer described in Blin

[4] Guerard (2012) decomposed the MQ variable into: (1) price momentum, (2) the consensus analysts' forecasts efficiency variable, CIBF, which itself is composed of forecasted earnings

et al. (1997) to create portfolios during the 1997–2009 period by varying the portfolio lambda. One seeks to maximize the Geometric Mean, Sharpe Ratios, and Information Ratios of portfolios. However, what if one wants to be considered a "concentrated portfolio manager" who does not hold 300–500 stocks. How many securities should one employ in portfolios using MV and EAW construction models with a monthly set of 3,000 expected return and covariance data? Can a manager construct efficient portfolios of 3,000 stock universes with fewer than 100 securities in the portfolios?

Guerard (2012) demonstrated the effectiveness of APT and Sungard APT systems in portfolio construction and management. Let us review the APT approach to portfolio construction. The estimation of security weights, x, in a portfolio is the primary calculation of Markowitz's portfolio management approach, as we have discussed in several chapters. The issue of security weights will be now considered from a different perspective. As previously discussed, the security weight is the proportion of the portfolio's market value invested in the individual security.

$$x_s = \frac{\mathrm{MV}_s}{\mathrm{MV}_\mathrm{p}}, \tag{7.9}$$

where x_s = portfolio-weight insecurity s, MV_s = value of security s within the portfolio, and MV_p = the total market value of portfolio.

The active weight of the security is calculated by subtracting the security weight in the (index) benchmark, b, from the security weight in the portfolio, p.

$$x_{s,\mathrm{p}} - x_{s,\mathrm{b}}. \tag{7.10}$$

Accordingly, if IBM has a 3% weight in the portfolio while its weight in the benchmark index is 2 and 1½ %, then IBM has a positive, 50 basis points active weight in the portfolio. The portfolio manager has an active, positive opinion of securities on which he or she has a positive active weight and a negative opinion of those securities with negative active weights.

---

yield, EP, revisions, EREV, and direction of revisions, EB, identified as breadth, Wheeler (1991), and (3) the stock standard deviation, identified in Malkiel (1963) as a variable with predictive power regarding the stock price-earnings multiple. Guerard (1997) and Guerard and Mark (2003) found that the consensus analysts' forecast variable dominated analysts' forecasted earnings yield, as measured by I/B/E/S 1-year-ahead forecasted earnings yield, FEP, revisions, and breadth. Guerard reported domestic (US) evidence that the predicted earnings yield is incorporated into the stock price through the earnings yield risk index. Moreover, CIBF dominates the historic low price-to-earnings effect, or high earnings-to-price, PE. We use CTEF, PRGR, and CIBF interchangeably in this monograph.

It is extremely important for the author to state that this section draws heavily from the APT Analytics Guide, written originally by John Blin and Steve Bender. The author's paper with Blin and Bender demonstrated how a statistically significant expected model can be used in the APT system to create portfolios that can produce excess returns (and statistically significant asset selection). The author is extremely grateful to APT and Sungard APT for its work with McKinley Capital Management.

Markowitz analysis (1952, 1959) and its efficient frontier minimized risk for a given level of return. Risk can be measured by a stock's volatility, or the standard deviation in the portfolio return over a forecast horizon, normally 1 year.

$$\sigma_{\mathrm{p}} = \sqrt{E(r_{\mathrm{p}} - E(r_{\mathrm{p}}))^2}. \tag{7.11}$$

Blin and Bender created an APT, Advanced Portfolio Technologies, Analytics Guide (2005 and 2011), which built upon the mathematical foundations of their APT system, published in Blin et al. (1997). The following analysis draws upon the APT analytics. Volatility can be broken down into systematic and specific risk:

$$\sigma_{\mathrm{p}}^2 = \sigma_{\beta\mathrm{p}}^2 + \sigma_{\varepsilon\mathrm{p}}^2, \tag{7.12}$$

where σ_{p} = total portfolio volatility, $\sigma_{\beta\mathrm{p}}$ = systematic portfolio volatility, and $\sigma_{\varepsilon\mathrm{p}}$ = specific portfolio volatility.

Blin and Bender created a multifactor risk model within their APT risk model based on forecast volatility.

$$\sigma_{\mathrm{p}} = \sqrt{52\left(\sum_{c=1}^{c}\left(\sum_{i=1}^{s} x_i\beta_{i,c}\right)^2 + \sum_{i=1}^{s} x_i^2\varepsilon_{i,x}^2\right)}, \tag{7.13}$$

where σ_{p} = forecast volatility of annual portfolio return, C = number of statistical components in the risk model, x_i = portfolio weight in security i, $\beta_{i,c}$ = the loading (beta) of security i on risk component c, and $\varepsilon_{i,\mathrm{w}}$ = weekly specific volatility of security i.

The Blin, Bender, and Guerard (1997) systematic volatility is a forecast of the annual portfolio standard deviation expressed as a function of each security's systematic APT components.

$$\sigma_{\beta\mathrm{p}} = \sqrt{52\sum_{c=1}^{c}\left(\sum_{i=1}^{s} x_i\beta_{i,c}\right)^2}. \tag{7.14}$$

Portfolio-specific volatility is a forecast of the annualized standard deviation associated with each security's specific return.

$$\sigma_{\varepsilon\mathrm{p}} = \sqrt{52\sum_{i=1}^{s} x_i^2\varepsilon_{i,x}^2}. \tag{7.15}$$

Tracking error, te, is a measure of volatility applied to the active return of funds (or portfolio strategies) indexed against a benchmark, which is often an index. Portfolio tracking error is defined as the standard deviation of the portfolio return less the benchmark return over 1 year.

$$\sigma_{te} = \sqrt{E(((r_p - r_b) - E(r_p - r_b))^2)}, \tag{7.16}$$

where σ_{te} = annualized tracking error, r_p = actual portfolio annual return, and r_b = actual benchmark annual return.

The APT-reported tracking error is the forecast tracking error for the current portfolio versus the current benchmark for the coming year.

$$\sigma_{te} = \sqrt{52\left(\sum_{c=1}^{C}\left(\sum_{i=1}^{S} x_{i,p} - x_{i,b}\right)\beta_{i,c}\right)^2 + \sum_{i=1}^{S}(x_{i,p} - x_{i,b})^2 \varepsilon_{i,x}^2}, \tag{7.17}$$

where $x_{i,p} - x_{i,b}$ = portfolio active weight.

Systematic Tracking Error of a portfolio is a forecast of the portfolio's active annual return as a function of the securities' returns associated with APT risk model components.

$$\sigma_{\beta te} = \sqrt{52\sum_{c=1}^{C}\left(\sum_{i=1}^{S}(x_{i,p} - x_{i,b})\beta_{i,c}^2\right)}. \tag{7.18}$$

Portfolio-specific tracking error can be written as a forecast of the annual portfolio active return associated with each security's specific behavior.

$$\sigma_{\varepsilon te} = \sqrt{52\sum_{i=1}^{S}(x_{i,p} - x_{i,b})^2 \varepsilon_{i,x}^2}. \tag{7.19}$$

The marginal volatility of a security, or the measure of the sensitivity of portfolio volatility, is relative to the change in the specific security weight.

$$\partial_s = \frac{\partial \sigma_p}{\partial x_s}, \tag{7.20}$$

where ∂_s = marginal risk of security s.

$$\partial_s = \beta_{sp}\sigma_p. \tag{7.21}$$

The portfolio Value-at-Risk (VaR) is the expected maximum loss that a portfolio could produce over 1 year.

$$\text{VaR} = \upsilon_p = \tilde{V}_T \text{ given prob}(V_T < \tilde{V}_T) = c, \tag{7.22}$$

where V_T = actual potential portfolio value in 1 year, $\tilde{V}_T$ = potential portfolio value in 1 year, and c = desired confidence level for VaR (i.e., 95%).

If a portfolio return is assumed to be generated from a normal distribution, then

$$\upsilon_{\mathrm{p}} = \sqrt{2}\mathrm{erf}^{-1}(2_x - 1)\sigma_{\mathrm{p}} V_0, \tag{7.23}$$

where $\mathrm{erf}^{-1}(x)$ = inverse error function and V_0 = current portfolio value.

The APT calculated VaR is written like this:

$$\upsilon_{\mathrm{p}} = \sqrt{2}\mathrm{erf}^{-1}(2_x - 1)\left(\sqrt{52\left(\sum\left(\sum x_i\beta_{i,c}\right)^2 + \sum x_i^2\varepsilon_{i,x}^2\right)}\right)V_0. \tag{7.24}$$

The APT measure of portfolio risk estimating the magnitude that the portfolio return may deviate from the benchmark return over 1 year is referred to as TaR, or Tracking-at-Risk™.

$$T_{\mathrm{p}}^{V} = \sqrt{\left(\frac{1}{\sqrt{1-x}}\sigma_s\right)^2 + \left(\sqrt{2}\mathrm{erf}^{-1}(x)\sigma_\varepsilon\right)^2}, \tag{7.25}$$

where T_{p}^{V} = TaR™, x = desired confidence level of TaR™, σ_s = portfolio Systematic Tracking Error, $\mathrm{erf}^{-1}(x)$ = inverse error function, and σ_ε = portfolio-specific tracking error.

Blin, Bender, and Guerard (1997) estimated a 20-factor beta model of covariances based on three-and-one-half years of weekly stock returns data. The Blin and Bender Arbitrage Pricing Theory (APT) model followed the Ross factor modeling theory, but Blin and Bender estimated betas from at least 20 orthogonal factors. Blin and Bender never sought to identify their factors with economic variables. Financial Economists, such as Phoebus Dhrymes and Mustafa Gultekin, often question the APT system for 20 orthogonal beta estimates. Dhrymes, Friend, and Gultekin (1983) and Dhrymes, Friend, Gultekin, and Gultekin (1984) found statistical evidence to support 4-5 factors. The estimation of 20 betas, when only 4-5, or at many as 7-8 in some periods, are needed is not a problem when one estimates orthogonal factors and exposures.

Guerard et al. (2010) found that the APT-TaR estimation procedure helped in creating 130/30 portfolios relative to traditional Markowitz mean–variance and equal active weighting portfolios. Guerard (2012) reported very similar results in construct equal active weighting (EAW2 with 2% deviations), mean–variance (MV with a 4% maximum weight) and Mean–Variance Tracking Error at Risk (MVTaR) portfolios for January 1997 to December 2009 using 8% monthly turnover, after the initial portfolio is created, and 150 basis points of transactions costs each way with USA and Global Expected Returns series. Comparing EAW, MV, and MV TaR provides support for the MVTaR procedure in the USA, as TaR maximizes the Geometric Mean, Sharpe Ratio, and Information Ratio relative to EAW and MV. In the global universe, MVTaR maximizes the Geometric Mean and Sharpe Ratio. EAW maximizes the Information Ratio in global markets over this time period.

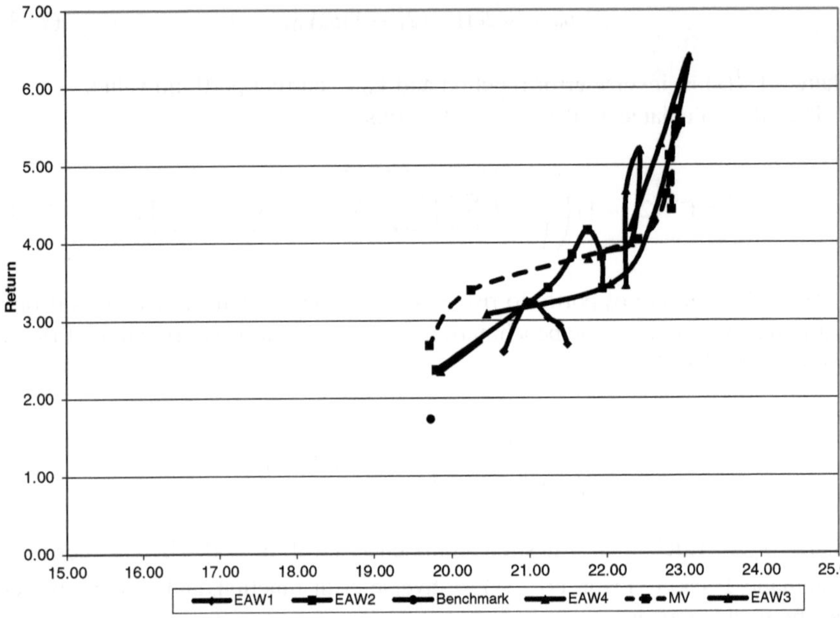

Fig. 7.1 USER Tracking Error at Risk (TaR) MV, EAW strategies, January 1997 to December 2009

Reported that APT-TaR estimation procedures were very successful in maximizing Information Ratios and Sharpe Ratios relative to MV and EAW techniques with the USER data.

Guerard, Krauklis, and Kumar (2012) reported that mean–variance dominated EAW1, EAW2, and EAW3 strategies with the USER data. One had to use an EAW4 to perform as well as mean–variance efficient frontier (Fig. 7.1).

The USER EAW1 curve showed no risk-return trade-off. An investor would be hard-pressed to outperform if he or she used an EAW1 strategy (unless he or she managed an index-enhanced product).

Guerard, Krauklis, and Kumar (2012) reported excess returns from APT-TaR portfolio results with the USER data. Let us review some of the Guerard, Krauklis, and Kumar (2012) results, shown in Table 7.1.

The Geometric Means, Sharpe Ratios, and Information Ratios for the mean–variance and Mean–Variance Tracking Error at Risk support the use of lambda equals 200 and the MVTaR approach.

It is well known that as one raises the portfolio lambda, the expected return of portfolio rises and the number of securities in the optimal portfolios fall, see Blin et al. (1997). Lambda, a measure of risk-aversion, the inverse of the risk-aversion acceptance level of the Barra system, is a decision variable to determine the optimal number of securities in a portfolio. As lambda rises, excess returns rise, Sharpe Ratios rise, and Information Ratios rise (as long as the expected returns model has a positive information coefficient, IC), and the number of securities in optimal portfolios fall. Guerard et al. (2010) report a lambda of 200 maximized the

Table 7.1 Optimal portfolio risk-return summary statistics

USER analysis							
January 1998 to December 2009							
	Mean–Variance (MV)				Mean–Variance Tracking Error at Risk (MVTaR)		
Lambda	Geometric Mean (GM)	Sharpe Ratio (ShR)	Information Ratio (IR)	Lambda	Geometric Mean (GM)	Sharpe Ratio (ShR)	Information Ratio (IR)
500	5.80	0.116	0.47	500	6.49	0.144	0.590
200	5.94	0.123	0.55	200	6.58	0.148	0.630
100	4.49	0.061	0.36	100	5.60	0.106	0.520
50	4.54	0.065	0.43	50	4.77	0.072	0.490
10	4.54	0.069	0.53	10	4.57	0.069	0.450
Benchmark	1.73	−0.074		Benchmark	1.73	−0.074	

Table 7.2 Average number of securities in optimal portfolios

USER analysis					
January 1997 to December 2009					
Lambda	EAW1	EAW2	EAW3	EAW4	MV
Tracking error at risk optimization					
500	118.1	85.4	74.7	68.6	64.8
200	122.6	92.2	82.5	77.4	77.8
100	125.6	100.0	92.2	90.5	90.5
50	131.3	111.7	105.0	103.5	103.4
10	147.3	137.4	133.7	133.2	136.2
Traditional optimization					
500	127.1	100.5	91.8	88.7	87.1
200	131.2	108.4	101.4	96.6	99.7
100	138.3	119.5	115.4	110.6	114.0
50	141.2	122.2	118.1	124.5	118.6
10	161.6	157.9	156.4	155.0	159.8

Geometric Mean in Non-USA growth portfolios. Guerard, Krauklis, and Kumar (2012) reported that the lambda of 200 is a necessary lambda with MV, EAW3, and EAW4 portfolio construction model for the USER data to create portfolios with fewer than 100 securities (Table 7.2).

Does the use of the TaR optimization technique produce a higher or lower number of average securities in portfolios than the MV optimization technique? A lambda of 200 implies optimal portfolios of 99.7 (100) stocks with mean–variance, MV, whereas MVTaR requires only 77.8 (78) stocks. The Blin and Bender TaR optimization procedure allows a manager to use fewer stocks in his or her portfolios than a traditional mean–variance optimization technique manager for a given lambda.

The reader notes that EAW1, EAW2, and EAW3 Tracking Error at Risk portfolios require more stocks than MVTaR and are statistically dominated in the risk-return trade-off curve, or Frontier, see Guerard, Krauklis, and Kumar (2012). In spite of the Markowitz mean–variance portfolio construction and management analysis being six decades old, it does very well in maximizing the Sharpe Ratio, Geometric Mean, and Information Ratio relative to newer approaches. The Markowitz Efficient Frontier (1952, 1956, 1959) methodology can be effectively implemented well with the USER data. Guerard, Krauklis, and Kumar (2012) reported that one must move to an EAW4 and EAW5 strategies to outperform Mean–Variance Tracking Error at Risk models. One cannot implement an EAWI strategy and "win" with a concentrated portfolio.

Portfolio Construction, Management, and Analysis: An Introduction to Systematic Tracking Error Optimization

It has been recognized for many years that sample covariance matrices are not the most suitable for portfolio optimization (Chopra and Ziemba (1993)). When the objective is to create a minimum variance portfolio, there are a series of shrinkage techniques which have been proposed to modify the sample covariance matrix $\boldsymbol{V}_{\text{sample}}$ (Ledoit and Wolf 2003, 2004), where the need for shrinkage is the estimation errors in the sample covariance matrix that may most likely render mean–variance optimizer less efficient. In its place, we suggest using the matrix obtained from the sample covariance matrix through a transformation called shrinkage. This tends to pull the most extreme coefficients towards more central values, systematically reducing estimation error. Wormald and van der Merwe (2012) searched, via shrinkage techniques, for a better estimate $\boldsymbol{V}_{\text{est}} \neq \boldsymbol{V}_{\text{sample}}$ for the covariance matrix to be used within an optimization, and in particular one which provides a more robust estimator of out-of-sample portfolio variances when used with quite general sets of expected return estimates.[5] Wormald and van der Merwe considered the advantages of using a factor model representation of the estimated covariance matrix $\boldsymbol{V}_{\text{est}}$ which appears in the Markowitz objective function for optimizing active (benchmark-relative) portfolios when expected returns (alphas) are available for every stock.

That Markowitz objective function takes the general form, in terms of the vector of weights w

$$U[w] = -\lambda \boldsymbol{w}^{\mathrm{T}} \cdot \boldsymbol{a} + {}^{1}\!/_{2} \boldsymbol{w}^{\mathrm{T}} \cdot \boldsymbol{V}_{\text{est}} \cdot \boldsymbol{w}, \tag{7.26}$$

where λ is called the risk trade-off parameter, and $\boldsymbol{a}$ is the vector of expected returns. A general factor decomposition of the covariance matrix $\boldsymbol{V}_{\text{est}}$ may be made in terms of asset exposures to factors $\boldsymbol{X}$, the covariance matrix $\boldsymbol{F}$ of the factors themselves, and the diagonal residual or specific term Δ^2

$$\boldsymbol{V}_{\text{est}} = \boldsymbol{V}_{\text{factor}} = \boldsymbol{X}^{\mathrm{T}} \boldsymbol{F} \boldsymbol{X} + \Delta^2. \tag{7.27}$$

The particular factor model representation we will consider is that provided by an orthonormalized Principal Components Analysis (PCA) factor model, such that the principal components factor covariances are all zero for different factors, and factor variances take the value 1.

Then we have, for these particular PCA factor exposures $\boldsymbol{B}$

$$\boldsymbol{F} = \boldsymbol{I}.$$

In this case, we can express the estimated asset covariance matrix in the special form

[5] There is a large literature on the application of optimization to portfolio construction, starting with Markowitz (1952, 1959) and reviewed in Fabozzi et al. (2002). A recent comprehensive overview can be found in the volume edited by Guerard (2010). An alternative approach might be pursued using ultrametrics and spanning trees rather than correlation shrinkage, see Onnela et al. (2003) for more on this approach.

$$\boldsymbol{V}_{\text{est}} = \boldsymbol{V}_{\text{PCA}} = \boldsymbol{B}^{\text{T}}\boldsymbol{B} + \Delta^2, \tag{7.28}$$

where $\boldsymbol{B}$ is the matrix of asset exposures to the orthonormalized factors and Δ^2 is the diagonal matrix of asset-specific risks in the model.

For Wormald and van der Merwe (2012), a key insight into the justification for factor modeling of risk is that it can be understood as an example of shrinkage techniques applied to the sample covariance matrix $\boldsymbol{V}_{\text{sample}}$, and has been widely accepted as an effective way of improving the risk characteristics of optimized portfolios, as described in Fabozzi, Jones, and Vardharaj (2002b). A parallel series of studies has focussed on the role of constraints in portfolio optimization, including contributions from Jagannathan and Ma (2003) and DeMiguel, Garlappi, Nogates, and Uppal (2008) who developed this line of inquiry and showed that many kinds of constraints applied in portfolio optimization can be understood as equivalent to statistically-sensible shrinkage of the sample covariance matrix. DeMiguel et al. (2008) focused on a detailed comparison of a set of portfolio strategies which are specified entirely by particular constraints defined in terms of the norm of the portfolio-weight vector, and provide a moment-shrinkage interpretation for the action of the constraint. In particular those authors prove analytically that quadratic constraints such as constraints on norms constructed from portfolio-weight vectors provide solutions which have a one-to-one correspondence with the portfolios proposed via the covariance shrinkage technique discussed in Ledoit and Wolf (2004). The empirical evidence they provide demonstrates that norm-constrained portfolios often have a higher Sharpe Ratio than less-constrained portfolio strategies and those considered by Jagannathan and Ma (2003) and Ledoit and Wolf (2003, 2004).

The issue of how best to apply shrinkage to the covariance matrix is also considered by Disatnik and Benninga (2007) who pay special attention to the use of shrinkage estimators and portfolios of estimators, a concept closely related to risk factor modeling. Their work, which is only concerned with the problem of constructing risk-minimized portfolios, suggests that short-sales constraints make a substantial difference in reducing the ex-post portfolio risk, compared to unconstrained global minimum solutions, and that it is quite difficult to obtain statistically significant differences from the ex-post risk for similarly-constrained solutions with differing covariance matrix estimators. This difficulty is one which also prevails when looking at the evidence for improved ex-post risk-adjusted performance when optimizing with an alpha model, which is the empirical case considered in the present study. When the objective is to create a portfolio with maximal alpha for a given risk (with either risk or alpha constrained to lie within bounds), there has been considerable attention paid to the question of whether the utility function should be modified to reflect the distinction between spanned and orthogonal alpha. The problem has been set out explicitly in Lee and Stefek (2008). The emphasis on treating spanned alpha (explained by the systematic factors of the risk model) and orthogonal alpha (not explained by those factors) differently within the utility function is motivated by very similar considerations to those treated in the literature on shrinkage approaches, where both the process of estimating expected asset return correlations via a model based on factors and the subsequent placing of constraints on portfolio norms (DeMiguel et al. 2008) have been shown to be effective in generating portfolios with significant out-of-sample improvement in risk characteristics.

Let us review the Wormald and van der Merwe (2012) distinctions between systematic and specific parts of the risk, since it is this distinction which underlies the concern that spanned alpha should be treated differently from orthogonal alpha within an optimization. The portfolio variance may in general be decomposed into a factor (systematic or spanned) part and a residual (specific or orthogonal) part:

$$\sigma_{\text{total}}^2 = \sigma_s^2 + \sigma_e^2. \tag{7.29}$$

The first part of the risk term, defined in terms of the portfolio-weight vector $\boldsymbol{w}$ as

$$\sigma_s^2 = \boldsymbol{w}^{\mathrm{T}} \cdot (\boldsymbol{B}^{\mathrm{T}}\boldsymbol{B}) \cdot \boldsymbol{w}, \tag{7.30}$$

the factor risk of the portfolio, while the second part of the risk term, defined as

$$\sigma_e^2 = \boldsymbol{w}^{\mathrm{T}} \cdot \Delta^2 \cdot \boldsymbol{w}, \tag{7.31}$$

the specific risk of the portfolio.

Wormald and van der Merwe (2012) demonstrated via the USER strategy simulation how the APT optimizer can be useful in implementing portfolio construction. Solutions which are constrained to be bounded on both systematic and specific risk terms require a second-order cone solver for efficient solutions, as described in Kolbert and Wormald (2010).

A great advantage in having efficient methods available to generate these solutions is that the investor's intuition can be tested and extended as the underlying utility or the investment constraints are varied. We present an analysis of the effects of the systematic risk constraint on various style exposures including momentum within the strategy simulation.

The objective function, to be minimized, for the optimization is now defined in terms of the *active* weight vector $\boldsymbol{w}$ of the portfolio, is given by exact analogy in:

$$U[\boldsymbol{w}] = -\lambda \boldsymbol{w}^{\mathrm{T}} \cdot \boldsymbol{a} + {}^1\!/_2 \boldsymbol{w}^{\mathrm{T}} \cdot (\boldsymbol{B}^{\mathrm{T}}\boldsymbol{B} + \Delta^2) \cdot \boldsymbol{w}, \tag{7.32}$$

where λ is the risk trade-off parameter and $\boldsymbol{a}$ is the vector of MQ alphas.

The covariance matrix is given by the APT factor model representation of (7.5):

$$V_{\text{pca}} = \boldsymbol{B}^{\mathrm{T}}\boldsymbol{B} + \Delta^2, \tag{7.33}$$

where $\boldsymbol{B}$ is the matrix of asset exposures to the APT factors and Δ^2 is the diagonal matrix of asset-specific risks in the model. In the empirical results set out here, we are concerned with active risk measures, and so we introduce the terminology of tracking error (TE) rather than variance for describing the factor and non-factor parts of the active risk. The effects of shrinkage in factor model estimation are demonstrated by considering the 2-part form of the total active risk (tracking error squared) term; we write, following the analogy with (7.32):

$$\sigma^2_{A\,\mathrm{total}} = \sigma^2_{As} + \sigma^2_{Ae}. \tag{7.34}$$

The first part of the risk term, defined as

$$\sigma^2_{As} = \boldsymbol{w}^{\mathrm{T}} \cdot (\boldsymbol{B}^{\mathrm{T}}\boldsymbol{B}) \cdot \boldsymbol{w}, \tag{7.35}$$

the active systematic risk (or systematic TE squared) of the portfolio, while the second part of the risk term, defined as

$$\sigma^2_{Ae} = \boldsymbol{w}^{\mathrm{T}} \cdot \Delta^2 \cdot \boldsymbol{w}, \tag{7.36}$$

the active specific risk (or specific TE squared) of the portfolio. Wormald and van der Merwe (2012) demonstrated the effects of shrinkage implied by optimization constraints within the empirical results, by putting separate constraints on the total TE and the systematic TE during the optimized USER simulations.

Wormald and van der Merwe (2012) implemented three strategies. The three strategies are very similar, except for differences in systematic active risk constraints. The first strategy constructs portfolios without any constraints on Systematic Tracking Error (TE), and is referred to as *NoRiskConst*. Another strategy places a mild constraint on systematic TE and is referred to as *MildRiskConst*. The mild constraint level reflects a level of systematic TE slightly lower than the average of the observed values in *NoRiskConst*. In *MildRiskConst* systematic TE is constrained to be below 2.3%. The third strategy constrains systematic TE to be below 1.5% and is called *StrongRiskConst*. Wormald and van der Merwe (2012) reported USER simulation results suggesting that applying a mild Systematic TE constraint leads to slight outperformance in the long run compared to other strategies. All three strategies outperform the benchmark. The Systematic Tracking Error methodology of Wormald and van der Merwe (2012) offered statistically significant asset selection and effective portfolio construction and management.

We use an All Country World Growth (ACWG) index and its constituents for the January 2002 to December 2011 period. We use a lambda of 200 and employ the Wormald and van der Merwe (2012) risk parameters. We find that the No Risk Control and Mild Risk Control simulations dominate the Strong Risk Control simulation, a result consistent with Wormald and van der Merwe. The three risk models work well, producing at least 700 basis points of outperformance, subtracting 150 basis points of transactions costs, each way, please see Table 7.3. The author highlights the work of Wormald and Van der Merwe because he believes that the TE risk constraints/controls represents a major enhancement within the APT system.

Table 7.3 Mild, strong, and no risk controls in a global universe, January 2002 to December 2011

Universe: All Country World Growth (ACWG)				
Model	Geometric Mean	Sharpe Ratio	Information Ratio	STD
No risk control	14.16	0.53	0.65	23.20
Mild risk control	13.75	0.49	0.59	24.18
Strong risk control	11.08	0.41	0.54	22.71
Benchmark	4.56	0.16		13.23

STD portfolio standard

Markowitz Restored: The Alpha Alignment Factor Approach

Several academics and practitioners, decided to perform a postmortem analysis of the mean–variance portfolios, attempted to understand the reasons for the deviation of ex-post performance from ex-ante targets and used their analysis to suggest enhancements to Markowitz's original approach. The reader may remember how the CAPM and beta was assailed in the early 1980s. The author took his Ph.D. seminar from Jan Mossin of Mossin (1966) and Mossin (1973) fame. Jan Mossin, when asked about the Roll criticism of the CAPM said "he is correct, but offers no solution." Criticism of one beta does not imply criticism of a multi-factor model. Lee and Stefek (2008) and Saxena and Stubbs (2012) have worked on optimization models to "restore" a better relationship between ex-ante and ex-post risk model estimates. One of the fundamental contributions was the development of linear factor models to capture the sources of systematic risk and characterize the key drivers of excess returns. While predicting expected returns is exclusively a forward looking activity, risk prediction also focuses on explaining cross-sectional variability of the returns process, mostly by using historical data. The first moment of the equity returns process drives expected return modelers while the second moment is the focus of risk modelers. These differences in the ultimate goals inevitably introduce certain "misalignment" between the factors used to forecast expected returns and risk. While expected return and risk models are indispensable components of any active strategy, there is a third component, namely, the set of constraints used to build a portfolio. Constraints play an important role in determining the composition of the optimal portfolio. Most real-life quantitative strategies have constraints that model desirable characteristics of the optimal portfolio. While some of these constraints may be mandatory, for example, a client's reluctance to invest in stocks that benefit from alcohol, tobacco or gambling activities on ethical grounds, other constraints are the result of best practices in practical portfolio management. A turnover constraint may create a factor misalignment, as we will find shortly in the USER analysis.

Saxena and Stubbs (2012) summarize, quantitative equity portfolio construction entails complex interaction between factors used for forecasting expected returns, risk, and the constraints. Problems that arise due to mutual discrepancies between these three entities are collectively referred to as Factor Alignment Problems (FAP) and constitute the emphasis of the current paper. Our key contributions are summarized below:

1. The differences in the approaches that are used to build expected return forecast and risk models manifest themselves as misalignment between the alpha and risk factors.
2. Using an optimization tool to construct the optimal holdings has the unintended effect of magnifying sources of misalignment. The optimize underestimates the systematic risk of the portion of the expected returns which is not aligned with the risk model. Consequently, it overloads the portion of the expected returns which is uncorrelated with all the user risk factors.
3. Our empirical results on a test-bed of real-life active portfolios based on client data clearly refute the validity of the assumption that the portion of alpha that is uncorrelated with all the risk factors has no systematic risk, and suggest the existence of systematic risk factors which are missing from the risk model.
4. We propose augmenting the risk model with an additional auxiliary factor to account for the effect of the missing risk factors in the risk model. The augmenting factor is constructed dynamically and takes a holistic view of the portfolio construction process involving the alpha model, the risk model, and the constraints. We provide analytical evidence to attest the effectiveness of the proposed approach.
5. Alternatively, the risk model can be augmented by adding the factors that are used to compute expected returns, and which are not represented in the risk model. The addition of these factors will provide full alignment between the risk model and the expected returns, but not necessarily handle any misalignment issues due to the use of constraints.

Quantitative strategies are typically based on three key components, namely, expected returns (or alphas), a risk model, and the constraints. The risk model is geared towards explaining cross-sectional variability in the historical and predicted returns. The efficacy of a risk model is judged by its ability to capture systematic risk factors and the correlation structure between their respective factor returns. The disparity in their respective objectives naturally affects the factors that are used in the linear models that are used in their construction, and introduces misalignment. With its primary focus on explaining the cross-sectional variability of the return process, a risk model can often make do with ballpark estimates and gains little, if at all, from razor sharp estimation of accounting entries. In the King's English, expected returns and risk modelers have different beliefs about the possible impact, or lack thereof, of various economic events on their respective mandates, and the misalignment between the alpha and risk factors is simply an inevitable manifestation of their diverse beliefs.

Second, expected returns and risk model developers can at times take a completely different view on the issue of earnings potential altogether. For instance, some alpha construction techniques use alternative valuation metrics such as different definitions of operating earnings and free cash flow for good reasons. These different measurement choices of the same underlying fundamental metric, namely earnings potential, lead to misalignment between the alpha and risk factors. Another source of misalignment arises from the use of book-to-price (B/P) ratio. Roughly speaking, book value is the accounting profession's estimate of the company's value; it reflects what the company paid for the assets except intangible assets such as goodwill developed internally, but it includes goodwill of subsidiary companies acquired by purchase. This "cost basis" is then adjusted downward by depreciation and amortization in a highly stylized and rigid attempt to reflect the economic depreciation that actually befalls (most) assets. Off balance-sheet items are ignored.

Saxena and Stubbs (2012) applied their AAF methodology to the USER model, running a monthly backtest based on the above strategy in 2001–2009 time period for various values of σ chosen from $\{0.5\%, 0.6\%, \ldots, 3.0\%\}$. For each value of σ, Saxena and Stubbs (2012) ran the backtest in two setups that were identical in all respects except one, namely, only the second setup used the AAF methodology (AAF = 20%). Saxena and Stubbs (2012) used Axioma's fundamental medium horizon risk model (US2AxiomaMH) to model the active risk constraint. Saxena and Stubbs (2012) reported the time series of the misalignment coefficient of alpha, implied alpha, and the optimal portfolio and found that almost 40–60% of the alpha is not aligned with the risk factors. The alignment characteristics of implied alpha are significantly better than that of alpha. Among other things, this implies that the constraints of the above strategy, especially the long-only constraint, play a proactive role in containing the misalignment issue. Saxena and Stubbs (2012) reported that the orthogonal component of both alpha and implied alpha not only has systematic risk but the magnitude of the systematic risk is comparable to the systematic risk associated with a median risk factor in US2AxiomaMH. To summarize, the primary purpose of portfolio optimization is to create a portfolio having an optimal risk-adjusted expected return. If a portion of the risk in a portfolio derived from the orthogonal component of implied alpha is not accounted for, then the resulting risk-adjusted expected return cannot be optimal. Saxena and Stubbs (2012) showed the predicted and realized active risk for various risk target levels, noting the significant downward bias in risk prediction when the AAF methodology is not employed.[6] Saxena and Stubbs (2012) showed the realized risk-return frontier and reported that using the AAF methodology not only improves the accuracy of risk prediction but also moves the ex-post frontier upwards thereby giving ex-post performance improvements. The distinguishing feature of quantitative investing as a profession is its belief in generating optimal risk-adjusted returns.

Saxena and Stubbs (2012) held that an approach that cannot predict the risk of the portfolio correctly cannot be expected to produce portfolios that are optimal in the ex-post sense. In other words, such an approach compromises the greater goal of

---

[6] The Bias statistic, shown is a statistical metric which is used to measure the accuracy of risk prediction; if the ex-ante risk prediction is unbiased, then the bias statistic should be close to 1.0 (see Saxena and Stubbs 2010 for more details). Clearly, the bias statistics obtained without the aid of the AAF methodology are significantly above the 95% confidence interval thereby showing that the downward bias in the risk prediction of optimized portfolios is statistically significant. The AAF methodology recognizes the possibility of inadequate systematic risk estimation and guides the optimizer to avoid taking excessive unintended bets.

Markowitz MV efficiency and yields suboptimal portfolios. The AAF approach, on the other hand, recognizes the possibility of missing systematic risk factors and makes amends to the extent possible without complete recalibration of the risk model that explicitly accounts for the latent systematic risk in alpha factors. In the process of doing so, AAF approach not only improves the accuracy of risk prediction but also partly repairs the lack of efficiency in the optimal portfolio.

Saxena and Stubbs (2012) acknowledged that the AAF approach has three key limitations. First, the AAF construct is based on the assumption that the factor returns associated with the missing factors are uncorrelated with the factor returns associated with the regular factors in the user risk model. The fact that the AAF is orthogonal to the regular factors, by itself, does not imply lack of correlation of factor returns. To see this, note that even though the industry factors derived from the GICS classification scheme are mutually orthogonal, the corresponding factor returns are often correlated. By being correlation agnostic, the AAF approach fails to capture the interaction between factor returns that can be attributed to missing factors and the user risk factors. Second, the AAF approach requires calibration of the volatility parameter which presents additional practical problems. Furthermore, the temporal stationarity of the mentioned volatility parameter is not guaranteed, which introduces additional complications related to dynamic estimation of the volatility parameter. Third, the AAF approach does not use historical data to improve its representation of the missing factors. In other words, it is agnostic to the nature of residual returns which might have useful information regarding missing factors. A natural way to circumvent these problems is to recalibrate the user risk model taking into account the possible sources of latent systematic risk. Saxena and Stubbs (2012) hold that Custom Risk Models (CRM) accomplish exactly that goal. CRM are derived from the user risk model, referred to as the base model, by introducing additional factors with the intent to eliminate various sources of misalignment. The additional factors are referred to as custom risk factors, and the resulting risk models are said to be customized. Construction of CRM involves complete recalibration of the covariance matrix by re-running the cross-sectional regressions, recomputing factor returns attributed to the user and custom risk factors, and using the resulting time series of factor and residual returns to compute the factor–factor covariance matrix and specific risk. To summarize, Saxena and Stubbs (2012) believe that a combination of CRM and AAF approach offers a practical and holistic approach to FAP. Saxena, Martin, and Stubbs (2011) developed a very readable Axioma Advisor, Research Focus, that complements the Saxena and Stubbs (2012) paper.

Let us take a final look at the USER data and portfolios using the Axioma Attrribution Analysis. If one uses the Axioma Medium Horizon Fundamental Risk Model for analyzing the APT-constructed ($\lambda = 200$) results reported in Guerard et al. (2012), one finds that asset selection dominates the portfolio returns; factor-based returns are -5.6 (%) whereas specific returns for 16.3%. The asset selection (active) of the APT-estimated USER model is 9.7% with an Information Ratio of 1.12 and a t-statistic of 3.68, see Table 7.4. The IRs and t-statistics are similar to those reported in Guerard et al. (2012). Furthermore, what about testing the USER

Table 7.4 Axioma Fundamental Risk Model attribution of APT Lambda = 200 Portfolio Returns (USER data, January 1999 to December 2009)

Total returns						
Portfolio						0.095
Benchmark						−0.012
Active						0.107

Local returns	Return	Risk	IR	T-Stat	Beg # of assets	End # of assets
Portfolio	0.095	0.221	n/a	n/a	94	92
Benchmark	−0.012	0.221	n/a	n/a	1854	1878
Active	0.107	0.096	1.115	3.683	1898	1922

Factor/specific contribution breakdown	
Factor contribution	−0.056
Specific return contribution	0.163
Active return	0.107

Return decomposition

Contributor	Return	Return	Return	Risk	IR	T-Stat
Risk-free rate	0.036					
Portfolio return	0.095					
Benchmark return	−0.012					
Active return	0.107			0.096	1.115	3.683
Market timing		0.000		n/a	n/a	n/a
Specific return		0.163		0.063	2.567	8.483
Factor contribution		−0.056		0.072	−0.778	−2.571
US2Axioma MH.Style			−0.045	0.068	−0.659	−2.178
US2Axioma MH. Industry			−0.011	0.032	−0.334	−1.103

	Contribution	Avg. Wtd. Exp.	HR	Risk	IR	T-Stat
Contributors to active return by US2AxiomaMH.Style						
US2AxiomaMH.Style						
US2AxiomaMH.Size	0.042	−1.053	0.557	0.047	0.881	2.911
US2AxiomaMH.Medium-Term Momentum	0.026	0.486	0.748	0.021	1.232	4.071
US2AxiomaMH.Value	0.010	0.433	0.710	0.008	1.286	4.248
US2AxiomaMH.Market Sensitivity	0.000	0.063	0.550	0.012	0.019	0.063
US2AxiomaMH.Exchange Rate Sensitivity	0.000	−0.357	0.580	0.007	0.027	0.090
US2AxiomaMH.Growth	−0.001	−0.044	0.443	0.002	−0.682	−2.252

(continued)

Table 7.4 (continued)

	Contribution	Avg. Wtd. Exp.	HR	Risk	IR	T-Stat
US2AxiomaMH.Short-Term Momentum	−0.007	0.055	0.405	0.010	−0.735	−2.428
US2AxiomaMH.Leverage	−0.011	0.351	0.458	0.008	−1.298	−4.288
US2AxiomaMH.Liquidity	−0.046	−1.148	0.351	0.036	−1.269	−4.194
US2AxiomaMH.Volatility	−0.057	0.399	0.244	0.022	−2.591	−8.560
US2AxiomaMH.Industry						
US2AxiomaMH.Computers & Peripherals	0.009	−0.047	0.473	0.016	0.528	1.744
US2AxiomaMH.Communications Equipment	0.008	−0.040	0.458	0.013	0.621	2.050
US2AxiomaMH.Pharmaceuticals	0.005	−0.062	0.496	0.016	0.307	1.014
US2AxiomaMH.Metals & Mining	0.004	0.022	0.618	0.010	0.343	1.135
US2AxiomaMH.Media	0.003	−0.006	0.565	0.005	0.758	2.505
US2AxiomaMH.Energy Equipment & Services	0.003	−0.017	0.473	0.007	0.485	1.603
US2AxiomaMH.Industrial Conglomerates	0.002	−0.044	0.450	0.012	0.193	0.637
US2AxiomaMH.Multiline Retail	0.002	−0.016	0.542	0.006	0.383	1.266
US2AxiomaMH.Food & Staples Retailing	0.002	−0.020	0.527	0.005	0.443	1.465
US2AxiomaMH.Specialty Retail	0.002	0.011	0.611	0.007	0.306	1.011
US2AxiomaMH.Aerospace & Defense	0.002	−0.013	0.450	0.005	0.436	1.441
US2AxiomaMH.Beverages	0.002	−0.020	0.504	0.006	0.335	1.108
US2AxiomaMH.Oil, Gas & Consumable Fuels	0.002	0.018	0.542	0.007	0.232	0.768
US2AxiomaMH.Machinery	0.002	−0.002	0.534	0.003	0.523	1.730
US2AxiomaMH.Household Products	0.002	−0.020	0.466	0.005	0.340	1.122
US2AxiomaMH.IT Services	0.001	−0.014	0.473	0.004	0.307	1.014
US2AxiomaMH.Tobacco	0.001	−0.008	0.489	0.003	0.292	0.966
US2AxiomaMH.Hotels, Restaurants & Leisure	0.001	−0.003	0.519	0.004	0.252	0.831
US2AxiomaMH.Electrical Equipment	0.001	−0.005	0.527	0.002	0.460	1.520
US2AxiomaMH.Biotechnology	0.001	0.014	0.527	0.006	0.128	0.424
US2AxiomaMH.Personal Products	0.001	−0.005	0.534	0.001	0.637	2.105
US2AxiomaMH.Road & Rail	0.001	−0.003	0.450	0.002	0.292	0.964
US2AxiomaMH.Independent Power Producers & Energy Traders	0.001	−0.002	0.443	0.001	0.589	1.947
US2AxiomaMH.Construction & Engineering	0.000	0.003	0.443	0.002	0.277	0.917
US2AxiomaMH.Diversified Consumer Services	0.000	0.000	0.489	0.002	0.135	0.447
US2AxiomaMH.Containers & Packaging	0.000	0.001	0.481	0.002	0.119	0.395
US2AxiomaMH.Gas Utilities	0.000	0.001	0.496	0.001	0.140	0.461

(continued)

Table 7.4 (continued)

	Contribution	Avg. Wtd. Exp.	HR	Risk	IR	T-Stat
US2AxiomaMH.Health Care Technology	0.000	0.000	0.382	0.000	0.674	2.227
US2AxiomaMH.Air Freight & Logistics	0.000	−0.006	0.473	0.002	0.058	0.190
US2AxiomaMH.Chemicals	0.000	0.007	0.573	0.003	0.030	0.099
US2AxiomaMH.Water Utilities	0.000	0.000	0.481	0.000	0.088	0.292
US2AxiomaMH.Transportation Infrastructure	0.000	0.000	0.076	0.000	0.481	1.589
US2AxiomaMH.Electric Utilities	0.000	0.009	0.496	0.005	−0.001	−0.004
US2AxiomaMH.Semiconductors & Semiconductor Equipment	0.000	−0.030	0.473	0.013	−0.006	−0.019
US2AxiomaMH.Office Electronics	0.000	−0.001	0.412	0.001	−0.171	−0.565
US2AxiomaMH.Consumer Finance	0.000	−0.005	0.481	0.004	−0.032	−0.107
US2AxiomaMH.Airlines	0.000	0.008	0.489	0.005	−0.028	−0.094
US2AxiomaMH.Construction Materials	0.000	0.000	0.489	0.001	−0.252	−0.833
US2AxiomaMH.Diversified Financial Services	0.000	−0.002	0.450	0.002	−0.139	−0.458
US2AxiomaMH.Food Products	0.000	0.008	0.565	0.004	−0.057	−0.189
US2AxiomaMH.Professional Services	0.000	−0.001	0.443	0.001	−0.346	−1.144
US2AxiomaMH.Distributors	0.000	0.001	0.527	0.000	−0.779	−2.575
US2AxiomaMH.Multi-Utilities	0.000	0.007	0.473	0.004	−0.130	−0.429
US2AxiomaMH.Software	−0.001	−0.033	0.450	0.010	−0.057	−0.190
US2AxiomaMH.Life Sciences Tools & Services	−0.001	−0.001	0.435	0.001	−0.538	−1.779
US2AxiomaMH.Building Products	−0.001	0.003	0.489	0.002	−0.265	−0.876
US2AxiomaMH.Thrifts & Mortgage Finance	−0.001	0.003	0.489	0.004	−0.151	−0.499
US2AxiomaMH.Marine	−0.001	0.000	0.389	0.000	−1.207	−3.990
US2AxiomaMH.Real Estate Investment Trusts (REITs)	−0.001	0.013	0.473	0.008	−0.080	−0.266
US2AxiomaMH.Health Care Equipment & Supplies	−0.001	0.002	0.443	0.004	−0.179	−0.591
US2AxiomaMH.Commercial Services & Supplies	−0.001	0.011	0.519	0.004	−0.180	−0.595
US2AxiomaMH.Leisure Equipment & Products	−0.001	0.002	0.405	0.002	−0.474	−1.565
US2AxiomaMH.Trading Companies & Distributors	−0.001	0.016	0.511	0.004	−0.223	−0.738
US2AxiomaMH.Capital Markets	−0.001	−0.006	0.443	0.004	−0.244	−0.805
US2AxiomaMH.Real Estate Management & Development	−0.001	0.004	0.504	0.001	−0.782	−2.583
US2AxiomaMH.Auto Components	−0.001	0.000	0.405	0.001	−1.115	−3.684
US2AxiomaMH.Health Care Providers & Services	−0.001	0.016	0.565	0.006	−0.211	−0.696

(continued)

Table 7.4 (continued)

	Contribution	Avg. Wtd. Exp.	HR	Risk	IR	T-Stat
US2AxiomaMH.Paper & Forest Products	−0.002	0.001	0.489	0.002	−0.995	−3.287
US2AxiomaMH.Commercial Banks	−0.002	0.037	0.496	0.014	−0.149	−0.492
US2AxiomaMH.Insurance	−0.003	0.013	0.427	0.007	−0.363	−1.199
US2AxiomaMH.Household Durables	−0.003	0.021	0.481	0.008	−0.366	−1.210
US2AxiomaMH.Textiles, Apparel & Luxury Goods	−0.003	0.025	0.511	0.008	−0.359	−1.187
US2AxiomaMH.Internet & Catalog Retail	−0.004	0.003	0.412	0.003	−1.363	−4.504
US2AxiomaMH.Automobiles	−0.004	0.023	0.458	0.008	−0.497	−1.643
US2AxiomaMH.Wireless Telecommunication Services	−0.004	0.022	0.511	0.006	−0.644	−2.126
US2AxiomaMH.Diversified Telecommunication Services	−0.006	0.049	0.489	0.012	−0.552	−1.824
US2AxiomaMH.Electronic Equipment, Instruments & Components	−0.009	0.039	0.534	0.014	−0.677	−2.237
US2AxiomaMH.Internet Software & Services	−0.016	0.018	0.405	0.013	−1.174	−3.880
US2AxiomaMH.Sectors						
US2AxiomaMH.Consumer Staples-S	0.007	−0.065	0.466	0.015	0.481	1.589
US2AxiomaMH.Energy-S	0.005	0.001	0.542	0.007	0.689	2.277
US2AxiomaMH.Industrials-S	0.005	−0.034	0.466	0.016	0.317	1.046
US2AxiomaMH.Health Care-S	0.003	−0.032	0.550	0.014	0.240	0.793
US2AxiomaMH.Materials-S	0.002	0.031	0.595	0.011	0.186	0.615
US2AxiomaMH.Utilities-S	0.000	0.015	0.565	0.006	0.041	0.135
US2AxiomaMH.Consumer Discretionary-S	−0.007	0.061	0.527	0.023	−0.294	−0.973
US2AxiomaMH.Information Technology-S	−0.008	−0.108	0.405	0.035	−0.225	−0.743
US2AxiomaMH.Financials-S	−0.009	0.058	0.466	0.024	−0.360	−1.190
US2AxiomaMH.Telecommunication Services-S	−0.010	0.071	0.496	0.016	−0.654	−2.160

model using higher targeting tracking errors in the Axioma system? We report, in Table 7.5, that the Geometric Means and Sharpe Ratios increase with higher targeted tracking errors while the Information Ratios fall (tracking errors increase more then realized portfolio returns) with USER in the Axioma system. The Geometric Means and Sharpe Ratios are higher in the Axioma 20-factor principal components estimated Statistical Risk Model than in the Axioma Fundamental Risk Model.

Table 7.5 The USER model with higher targeted tracking errors

USER model

Universe: R3000

Simulation period: January 1999 to March 2009

Transactions costs: 125 basis points each way

			NoAAF					AAF				
Return model	Risk model	Tracking Error	Sharpe Ratio	Information Ratio	Ann. active return (%)	Ann. active risk (%)	*N*	Sharpe Ratio	Information Ratio	Ann. active return (%)	Ann. active risk (%)	*N*
USER	STAT	4	0.48	1.83	13.26	7.24	160	0.48	2.20	12.81	5.83	241
		5	0.52	1.72	14.71	8.57	130	0.52	1.98	14.22	7.16	190
		6	0.52	1.56	15.27	9.77	110	0.54	1.79	15.10	8.45	157
		7	0.55	1.52	16.47	10.84	97	0.55	1.64	15.93	9.72	131
		8	0.55	1.39	16.89	12.13	86	0.58	1.58	17.08	10.82	110
	FUND	4	0.40	2.03	11.12	5.49	139	0.42	2.44	11.48	4.70	215
		5	0.45	1.89	12.53	6.64	119	0.46	2.15	12.66	5.89	170
		6	0.48	1.75	13.59	7.78	106	0.49	1.91	13.53	7.08	144
		7	0.48	1.56	13.95	8.93	96	0.50	1.73	14.19	8.19	123
		8	0.47	1.38	14.25	10.31	89	0.50	1.58	14.71	9.33	107

Table 7.6 Global composite model component ICs

January 1990 to September 2009	
Variable	IC
EP	0.048
BP	0.019
CP	0.042
SP	0.008
DP	0.058
RV1	0.011
RV2	0.019
BR1	0.026
BR2	0.024
FEP1	0.034
FEP2	0.029
CTEF	0.023
PM	0.022
EWC	0.043
GLER	0.042

An Global Expected Returns Model: Why Everyone Should Diversify Globally, 1998–2009

Guerard et al. (2012) extended a stock selection model originally developed and estimated in Guerard and Takano (1991) and Bloch, Guerard, Markowitz, Todd, and Xu (1993), adding a Brush-based price momentum variable, taking the price at time $t - 1$ divided by the price 12 months ago, $t - 12$, denoted PM, and the consensus (I/B/E/S) analysts' earnings forecasts and analysts' revisions composite variable, CTEF, to the stock selection model. Guerard et al. (2012) referred to the stock selection model as a United States Expected Returns (USER) model. We can estimate an expanded stock selection model to use as an input of expected returns in an optimization analysis. The universe for all analysis consists of all securities on Wharton Research Data Services (WRDS) platform from which we download the I/B/E/S database, and the Global Compustat databases. The I/B/E/S database contains consensus analysts' earnings per share forecast data and the Global Compustat database contains fundamental data, i.e., the earnings, book value, cash flow, depreciation, and sales data, used in this analysis for the January 1990 to December 2009 time period. The information coefficient, IC, is estimated as the slope of a regression line in which ranked subsequent returns are expressed as a function of the ranked strategy, at a particular point of time. The high fundamental variables, earnings, bookvalue, cash flow, and sales produce higher ICs in the global universe than in the USA universe where USER was estimated, see Table 7.6. Moreover, analysts' 1-year-ahead and 2-year ahead revisions, RV1 and RV2, respectively, were much lower in global markets, than USA market. Breadth, BR, and forecasted earnings yields, FEP, were positive but less than in the USA market. The ICs on the analysts' forecast variable, CTEF, and price momentum variable, PM, were lower than in the (domestic) USA universe.

The stock selection model estimated in this study, denoted as Global Expected Returns, GLER, is:

$$\begin{aligned}\mathrm{TR}_{t+1} = a_0 + a_1\mathrm{EP}_t + a_2\mathrm{BP}_t + a_3\mathrm{CP}_t + a_4\mathrm{SP}_t + a_5\mathrm{REP}_t + a_6\mathrm{RBP}_t \\ + a_7\mathrm{RCP}_t + a_8\mathrm{RSP}_t + a_9\mathrm{CTEF}_t + a_{10}\mathrm{PM}_t + e_t, \quad (7.37)\end{aligned}$$

where EP = [earnings per share]/[price per share] = earnings–price ratio; BP = [book value per share]/[price per share] = book–price ratio; CP = [cash flow per share]/[price per share] = cash flow–price ratio; SP = [net sales per share]/[price per share] = sales–price ratio; REP = [current EP ratio]/[average EP ratio over the past 5 years]; RBP = [current BP ratio]/[average BP ratio over the past 5 years]; RCP = [current CP ratio]/[average CP ratio over the past 5 years]; RSP = [current SP ratio]/[average SP ratio over the past 5 years]; CTEF = consensus earnings per share I/B/E/S forecast, revisions and breadth; PM = price momentum; and e = randomly distributed error term.

The GLER model also is estimated using a weighted latent root regression, WLRR, analysis on (7.1) to identify variables statistically significant at the 10% level; uses the normalized coefficients as weights; and averages the variable weights over the past 12 months. The 12-month smoothing is consistent with the four-quarter smoothing in Guerard and Takano (1991) and Bloch et al. (1993). While EP and BP variables are significant in explaining returns, the majority of the forecast performance is attributable to other model variables, namely the relative earnings-to-price, relative cash-to-price, relative sales-to-price, price momentum, and earnings forecast variables. The consensus earnings forecasting variable, CTEF, and the price momentum variable, PM, dominate the composite model, as is suggested by the fact that the variables account for 48% of the model average weights, slightly higher than the two variables combining for 44% of the weights in the USER model. The time-average value of estimated coefficients:

a_1	a_2	a_3	a_4	a_5	a_6	a_7	a_8	a_9	a_{10}
0.048	0.069	0.044	0.047	0.050	0.032	0.039	0.086	0.216	0.257.

In terms of information coefficients, ICs, the use of the WLRR procedure produces a virtually identical IC for the models during the 1980–2009 time period, 0.042, versus the equally-weighted IC of 0.043. The GLER model, has compared to the USER model in Guerard et al. (2012) has approximately the same ICs. The IC test of statistical significance can be referred to as a Level I test. Further evidence on the anomalies is found in Levy (1999).

We report that in the Axioma GLER simulations, as with USER, the Axioma Statistical Model dominates the Axioma Fundamental Model and AAF dominates the non-AAF Frontiers in terms of Geometric Means and Sharpe Ratios with the GLER Model (see Table 7.7).[7] Moreover, in Table 7.8, lower turnover (4%,

[7] The author worked on the GLER analysis with Anureet Saxena. Any errors remaining in this section are the sole responsibility of the author.

Table 7.7 An AAF analysis of the Global Expected Returns (GLER) model

Initial Axioma WRDS GLER Backtest

GLER model—global variation of USER

Universe: ACWG

Simulation period: January 1999 to March 2009

Transactions costs: 150 basis points each way, respectively

			No AAF					AAF				
Return model	Risk model	Tracking Error	Sharpe Ratio	Information Ratio	Ann. active return	Ann. active risk	*N*	Sharpe Ratio	Information Ratio	Ann. active return	Ann. active risk	*N*
GLER	STAT	4	0.448	1.247	8.72	6.99	216	0.290	1.159	4.79	4.14	516
		5	0.511	1.119	10.52	8.77	204	0.359	1.230	6.37	5.18	442
		6	0.516	1.089	11.02	10.12	188	0.397	1.145	7.43	6.49	383
		7	0.552	1.074	12.29	11.44	185	0.464	1.179	9.09	7.71	340
		8	0.605	1.111	14.14	12.73	177	0.532	1.236	10.94	8.86	304
	FUND	4	0.286	0.882	4.97	5.63	221	0.230	1.009	3.53	3.50	488
		5	0.320	0.841	5.84	6.94	199	0.269	0.971	4.45	4.59	414
		6	0.356	0.827	6.91	6.91	196	0.306	0.952	5.39	5.66	357
		7	0.414	0.885	8.45	8.45	188	0.344	0.946	6.36	6.72	318
		8	0.427	0.845	8.99	8.99	182	0.407	1.012	7.97	7.88	291

Table 7.8 An AAF/turnover analysis of the Global Expected Returns (GLER) model

Initial Axioma WRDS GLER Backtest																
GLER model—global variation of USER																
Universe: ACWG																
Axioma Statistical Risk Model																
Simulation period: January 1999 to December 2011																
Transactions costs: 150 basis points each way, respectively																
		AAF = 10					AAF = 30					AAF = 70				
Return model	Tracking Error	Sharpe Ratio	Information Ratio	Ann. active return	Ann. active risk	*N*	Sharpe Ratio	Information Ratio	Ann. active return	Ann. active risk	*N*	Sharpe Ratio	Information Ratio	Ann. active return	Ann. active risk	*N*
Combo10F																
TO = 4	4	0.490	1.491	9.37	6.29	444	0.413	1.407	7.64	5.43	334	0.287	1.289	4.29	3.66	601
	5	0.531	1.291	10.68	8.19	522	0.479	1.386	9.35	6.74	289	0.355	1.335	6.24	4.71	504
	6	0.569	1.217	11.90	9.84	637	0.519	1.267	10.61	8.37	526	0.397	1.261	7.37	5.84	439
	7	0.597	1.187	12.87	8.19	214	0.597	1.304	12.67	9.72	598	0.476	1.310	9.33	7.12	387
	8	DNC					0.609	1.249	13.50	10.81	229	0.531	1.315	10.84	8.25	347
TO = 8	4	0.426	1.215	8.14	6.71	228	0.353	1.129	6.39	5.66	294	0.279	1.203	4.53	3.77	607
	5	0.475	1.146	9.65	8.43	203	0.419	1.147	7.96	6.94	243	0.342	1.241	5.95	4.79	513
	6	0.519	1.104	11.07	10.03	193	0.494	1.178	9.88	8.38	214	0.393	1.232	7.24	5.87	446
	7	0.555	1.084	12.31	11.35	195	0.554	1.182	11.59	9.80	187	0.435	1.181	8.42	7.13	389
	8	0.615	1.141	13.94	12.22	401	0.591	1.193	13.12	11.00	185	0.501	1.222	10.06	8.32	359
TO = 12	4	0.413	1.142	7.86	6.88	211	0.351	1.122	6.35	5.65	294	0.263	1.198	4.22	3.77	538
	5	0.476	1.147	9.47	8.25	177	0.418	1.141	7.96	6.98	243	0.328	1.170	5.67	4.85	468
	6	0.554	1.192	11.51	9.66	163	0.493	1.175	9.82	8.36	214	0.381	1.177	6.95	5.91	481
	7	0.581	1.176	12..82	10.90	162	0.559	1.198	11.72	9.78	192	0.424	1.147	8.11	7.07	366
	8	0.622	1.157	14.16	12.24	171	0.601	1.197	13.22	11.85	82	0.459	1.002	9.09	8.25	335

DNC did not converge

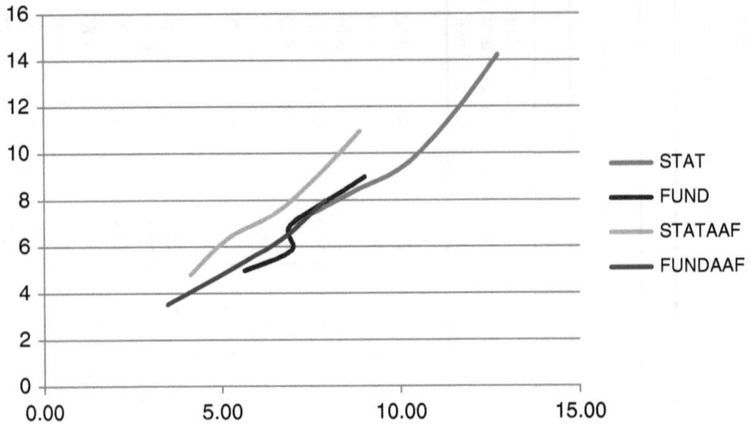

Fig. 7.2 Dominance of the Statistical Risk Model and Alpha Alignment Factors relative to Fundamental Risk Models and No Alpha Alignment Factors in the United States Equity Market, 1999–2009

monthly) allows the AAF factor to increase. An AAF of 30% is preferred to AAF levels of 10 or 70%, for most tracking errors and turnover. The GLER model risk-return frontier demonstrates the effectiveness of the USER analysis in global markets. Finally, if one graphs portfolio excess returns relative to portfolio tracking errors, one sees in Fig. 7.2 that the Axioma Statistical Risk Model frontier with AAF = 30% dominates the Axioma Statistical Risk Model frontier without AAF. Furthermore, the Axioma Statistical Risk Model frontier dominates the Axioma Fundamental Risk Model frontier (with and without AAF).

Global Investing in the World of Business, 1999–2011

In the world of business, one does not access academic databases annually, or even quarterly. Most industry analysis uses FactSet database and the Thomson Financial (I/B/E/S) earnings forecasting database. We can estimate (7.37) for all securities on the Thomson Financial and FactSet databases, some 46,550 firms in December 2011. We can decompose this universe into USA, Non-USA, and global securities. We can refer to these universes as the USER, NUSER, and GLER databases, respectively. One can estimate (7.37) models for index constituents in the three growth universes: the Russell 3000 Growth (R3G) universe; the MSCI All Country World ex USA Growth (ACWexUSG) universe; and the All Country World Growth (ACWG) universe. The R3G analysis is shown in Table 7.9; the ACWexUSG analysis is reported in Table 7.10; and ACWG universe analysis is shown in Table 7.11. The GLER conclusions are confirmed: (1) the Axioma Statistical Model dominates the Axioma Fundamental Model and (2) AAF dominates the

Table 7.9 Axioma analysis of Russell 3000 growth constituents

Initial Axioma Ranked USA Backtest

USER model

Universe: R3G

Simulation period: January 1999 to December 2011

Transactions costs: 150 basis points each way, respectively

			NoAAF					AAF				
Return model	Risk model	Tracking Error	Sharpe Ratio	Information Ratio	Ann. active return	Ann. active risk	*N*	Sharpe Ratio	Information Ratio	Ann. active return	Ann. active risk	*N*
USER model	STAT	4	0.303	0.734	5.49	7.49	177	0.300	0.842	5.20	6.18	317
		5	0.309	0.650	5.78	8.90	138	0.336	0.835	6.07	7.29	242
		6	0.288	0.548	5.50	10.05	109	0.343	0.767	6.41	8.40	192
		7	0.301	0.536	6.04	11.25	91	0.345	0.700	6.70	9.59	156
		8	0.338	0.586	7.12	12.14	77	0.345	0.640	6.83	10.64	128
	FUND	4	0.239	0.643	4.09	6.35	188	0.252	0.767	4.27	5.55	323
		5	0.276	0.646	5.01	7.75	148	0.275	0.715	4.84	6.77	257
		6	0.311	0.657	5.96	9.08	122	0.305	0.720	5.65	7.84	205
		7	0.298	0.598	5.88	10.15	106	0.305	0.644	5.81	9.03	202
		8	0.301	0.547	6.16	11.26	91	0.327	0.645	6.49	10.07	140

Table 7.10 Axioma analysis of all ex USA growth index constituents

Initial Axioma Ranked NUSG Backtest

NUSER model—Non-USA variation of USER

Universe: ACWexUSG

Simulation period: January 1999 to December 2011

Transactions costs: 150 basis points each way, respectively

			RANKED									
			NoAAF					AAF				
Return model	Risk model	Tracking Error	Sharpe Ratio	Information Ratio	Ann. active return	Ann. active risk	*N*	Sharpe Ratio	Information Ratio	Ann. active return	Ann. active risk	*N*
NUSER	STAT	4	0.487	1.245	8.15	6.55	133	0.446	1.380	7.01	5.08	242
		5	0.546	1.228	9.79	7.97	102	0.501	1.377	8.43	6.12	182
		6	0.679	1.471	13.07	8.88	81	0.537	1.312	9.48	7.22	140
		7	0.719	1.450	14.53	10.02	66	0.618	1.394	11.56	8.29	113
		8	0.782	1.514	16.41	10.84	55	0.718	1.538	14.22	9.24	91
	FUND	4	0.445	1.271	6.68	5.25	153	0.405	1.331	5.79	4.35	244
		5	0.473	1.133	7.57	6.68	118	0.470	1.351	7.25	5.37	240
		6	0.557	1.232	9.66	7.84	95	0.518	1.340	8.42	6.43	152
		7	0.652	1.378	11.99	8.70	78	0.572	1.309	9.83	7.51	121
		8	0.725	1.465	14.06	9.60	66	0.640	1.374	11.74	8.54	100

Table 7.11 Axioma analysis of all country world growth index constituents

Initial Axioma Ranked Global Backtest

GLER model—global variation of USER

Universe: ACWG

Simulation period: January 1999 to December 2011

Transactions costs: 150 basis points each way, respectively

			RANKED									
			NoAAF					AAF				
Return model	Risk model	Tracking Error	Sharpe Ratio	Information Ratio	Ann. active return	Ann. active risk	*N*	Sharpe Ratio	Information Ratio	Ann. active return	Ann. active risk	*N*
GLER	STAT	4	0.554	1.475	9.99	6.78	144	0.489	1.507	8.51	5.65	261
		5	0.602	1.385	11.38	8.24	110	0.554	1.521	10.09	6.63	197
		6	0.656	1.409	13.25	9.40	87	0.614	1.502	11.65	7.76	153
		7	0.715	1.454	14.94	10.28	70	0.638	1.415	12.63	8.93	120
		8	0.748	1.451	16.20	11.16	58	0.672	1.385	14.00	10.11	95
	FUND	4	0.382	1.091	6.08	5.57	163	0.373	1.231	5.82	4.73	272
		5	0.460	1.151	7.73	6.72	129	0.438	1.260	7.19	5.71	210
		6	0.521	1.158	9.33	8.06	104	0.492	1.255	8.40	6.69	167
		7	0.582	1.217	11.02	9.06	83	0.563	1.294	10.08	7.79	137
		8	0.647	1.281	12.75	9.95	71	0.602	1.265	11.36	8.99	110

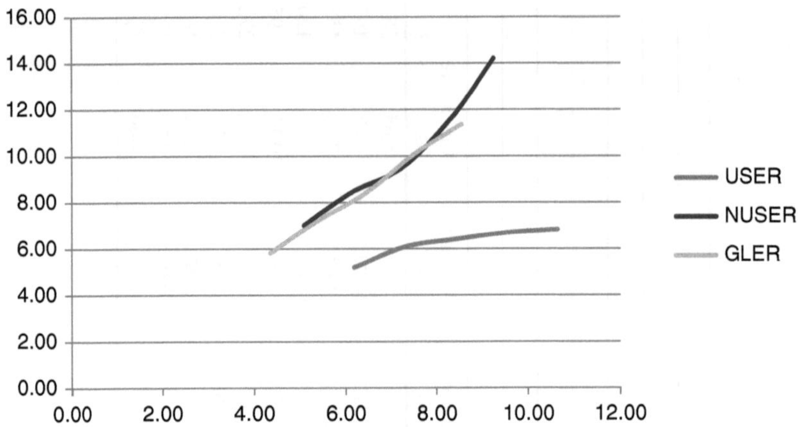

Fig. 7.3 It paid to be an International Investor, 1999–2011

Table 7.12 Portfolio criteria for no risk constraints STE portfolios

	Geometric Means	Sharpe Ratios	Information Ratios	Tracking Errors
Lambda = 500	17.96	0.67	0.92	14.69
Lambda = 200	14.16	0.53	0.65	14.63
ACWG benchmark	4.56	0.16		

non-AAF Frontiers in terms of Information Ratios and Sharpe Ratios with the models.[8] An examination of Tables 7.9, 7.10, and 7.11, shows that Non-USA and global models produce higher Sharpe Ratios and higher Information Ratios than the USER model in the 1999–2011 period. Non-USA and global stocks are more inefficient than USA stocks, a result reported in Guerard (2012). If we graph the USER, NUSER, and GLER active risks and active returns, we find that GLER and NUSER dominate USER (see Fig. 7.3). NUSER dominates GLER at an 8% tracking error.

Let us take a closer look at the application at the Systematic Tracking Error (STE) optimization technique reported in Wormald and van der Merwe (2012). Let us take the FactSet and Thomson Financial universe for the 1990–2011 period and reduce it by requiring at least two analysts to cover stocks. The

[8] It is interesting to note that initial Axioma analysis suggests that purchasing AWCG constituents produce similar Information Ratios and Sharpe Ratios to purchasing FactSet and Thomson Financial securities (with at least two analysts covering the stocks, a universe exceeding index constituents by a factor of 5–6 times). The similar Sharpe Ratios and IRs are very interesting given the very illiquid composition of many securities (trading volume of less than $15 MM USD, daily).

Table 7.13 Portfolio GLER L500 results

Portfolio	VA_NoRCGLER_McKinley
Benchmark	MSCI WORLD GROWTH
Attribution period	01/31/2003 to 01/31/2012
Frequency	Monthly
Risk model	WW21AxiomaMH
Bayesian half life	2.0
Realized market return (1/year)	0
Return type	Geometric
Risk scaling	Annualized
Risk type	PREDICTED_RISK
Report date	06/29/2012
Base currency	USD

Total returns	
Portfolio	0.187
Benchmark	0.080
Active	0.107

Local returns	Return	Risk	IR	T-Stat	Beg # of assets	End # of assets
Portfolio	0.187	0.232	n/a	n/a	108	123
Benchmark	0.080	0.192	n/a	n/a	528	965
Active	0.107	0.093	1.148	3.443	633	1075

Table 7.13 (continued)

Factor/specific contribution breakdown	
Factor contribution	0.042
Specific return contribution	0.065
Active return	0.107

Contributor	Return	Return	Return	Risk	IR	T-Stat
Return decomposition						
Risk-free rate	0.024					
Portfolio return	0.187					
Benchmark return	0.080					
Active return	0.107			0.093	1.148	3.443
Market timing		0.000		n/a	n/a	n/a
Specific return		0.065		0.060	1.082	3.246
Factor contribution		0.042		0.071	0.588	1.763
WW21AxiomaMH.Style			0.014	0.059	0.233	0.698
WW21AxiomaMH.Market			0.000	0.000	0.558	1.673
WW21AxiomaMH.Local			0.002	0.007	0.285	0.855
WW21AxiomaMH.Industry			0.001	0.017	0.066	0.198
WW21AxiomaMH.Currency			0.009	0.016	0.571	1.712
WW21AxiomaMH.Country			0.015	0.025	0.600	1.799

	Contribution	Avg. Wtd. Exp.	HR	Risk	IR	T-Stat
WW21AxiomaMH.Style						
WW21AxiomaMH.Medium-Term Momentum	0.045	0.321	0.741	0.017	2.573	7.718

(continued)

Table 7.13 (continued)

	Contribution	Avg. Wtd. Exp.	HR	Risk	IR	T-Stat
WW21AxiomaMH.Size	0.014	−0.894	0.565	0.044	0.324	0.972
WW21AxiomaMH.Value	0.010	0.436	0.620	0.011	0.898	2.694
WW21AxiomaMH.Liquidity	0.003	0.137	0.556	0.005	0.688	2.065
WW21AxiomaMH.Growth	0.003	0.314	0.583	0.004	0.783	2.348
WW21AxiomaMH.Exchange Rate Sensitivity	−0.001	0.083	0.444	0.002	−0.269	−0.806
WW21AxiomaMH.Leverage	−0.004	0.090	0.426	0.002	−2.312	−6.935
WW21AxiomaMH.Short-Term Momentum	−0.014	0.106	0.380	0.012	−1.183	−3.550
WW21AxiomaMH.Volatility	−0.043	0.630	0.380	0.049	−0.879	−2.636
Contributors to Active Return by WW21AxiomaMH.Market						
WW21AxiomaMH.Market						
WW21AxiomaMH.Global Market	0.000	0.000	0.602	0.000	0.558	1.673
Contributors to Active Return by WW21AxiomaMH.Local						
WW21AxiomaMH.Local						
WW21AxiomaMH.Domestic China	0.002	0.009	0.222	0.007	0.285	0.855
Contributors to Active Return by WW21AxiomaMH.Industry						
WW21AxiomaMH.Industry						
WW21AxiomaMH.Metals & Mining	0.004	0.043	0.556	0.006	0.632	1.897
WW21AxiomaMH.Media	0.002	−0.013	0.630	0.001	1.859	5.577
WW21AxiomaMH.Real Estate Investment Trusts (REITs)	0.002	0.026	0.528	0.003	0.567	1.701
WW21AxiomaMH.Pharmaceuticals	0.002	−0.052	0.481	0.005	0.334	1.003
WW21AxiomaMH.Communications Equipment	0.001	−0.018	0.519	0.002	0.507	1.521
WW21AxiomaMH.Wireless Telecommunication Services	0.001	0.031	0.565	0.002	0.497	1.492
WW21AxiomaMH.Health Care Providers & Services	0.001	0.008	0.537	0.003	0.291	0.874
WW21AxiomaMH.Thrifts & Mortgage Finance	0.001	0.001	0.481	0.001	0.974	2.923
WW21AxiomaMH.Transportation Infrastructure	0.001	0.009	0.509	0.001	0.869	2.607
WW21AxiomaMH.Internet & Catalog Retail	0.001	0.001	0.500	0.001	0.580	1.741
WW21AxiomaMH.Internet Software & Services	0.001	−0.003	0.537	0.001	0.594	1.782
WW21AxiomaMH.Electronic Equipment, Instruments & Components	0.000	−0.006	0.519	0.001	0.793	2.380
WW21AxiomaMH.Consumer Finance	0.000	−0.001	0.519	0.001	0.547	1.642
WW21AxiomaMH.Diversified Telecommunication Services	0.000	0.022	0.481	0.002	0.157	0.471
WW21AxiomaMH.Containers & Packaging	0.000	0.003	0.574	0.001	0.533	1.598

(continued)

Table 7.13 (continued)

	Contribution	Avg. Wtd. Exp.	HR	Risk	IR	T-Stat
WW21AxiomaMH.Professional Services	0.000	0.001	0.537	0.001	0.432	1.297
WW21AxiomaMH.Health Care Technology	0.000	0.007	0.315	0.001	0.254	0.761
WW21AxiomaMH.Aerospace & Defense	0.000	−0.007	0.481	0.001	0.180	0.539
WW21AxiomaMH.Commercial Banks	0.000	0.003	0.472	0.001	0.161	0.484
WW21AxiomaMH.Office Electronics	0.000	−0.005	0.528	0.000	0.420	1.260
WW21AxiomaMH.Hotels, Restaurants & Leisure	0.000	0.005	0.583	0.001	0.165	0.494
WW21AxiomaMH.Software	0.000	−0.033	0.463	0.003	0.053	0.159
WW21AxiomaMH.Computers & Peripherals	0.000	−0.017	0.463	0.002	0.057	0.171
WW21AxiomaMH.Textiles, Apparel & Luxury Goods	0.000	−0.002	0.593	0.001	0.215	0.646
WW21AxiomaMH.Construction & Engineering	0.000	−0.004	0.472	0.000	0.296	0.889
WW21AxiomaMH.Food Products	0.000	−0.010	0.491	0.001	0.103	0.308
WW21AxiomaMH.Construction Materials	0.000	0.000	0.509	0.000	0.138	0.414
WW21AxiomaMH.Multiline Retail	0.000	0.001	0.472	0.001	0.039	0.116
WW21AxiomaMH.Air Freight & Logistics	0.000	−0.008	0.519	0.001	0.048	0.143
WW21AxiomaMH.Water Utilities	0.000	0.001	0.380	0.000	0.012	0.037
WW21AxiomaMH.Household Durables	0.000	0.002	0.509	0.001	0.005	0.016
WW21AxiomaMH.Semiconductors & Semiconductor Equipment	0.000	−0.011	0.509	0.002	−0.018	−0.054
WW21AxiomaMH.Building Products	0.000	0.004	0.500	0.001	−0.121	−0.363
WW21AxiomaMH.Leisure Equipment & Products	0.000	0.000	0.546	0.000	−0.246	−0.739
WW21AxiomaMH.Electrical Equipment	0.000	−0.007	0.481	0.000	−0.189	−0.568
WW21AxiomaMH.Trading Companies & Distributors	0.000	0.002	0.435	0.000	−0.244	−0.733
WW21AxiomaMH.Independent Power Producers & Energy Traders	0.000	0.002	0.417	0.000	−0.267	−0.802
WW21AxiomaMH.Diversified Consumer Services	0.000	−0.001	0.509	0.000	−0.389	−1.167
WW21AxiomaMH.Industrial Conglomerates	0.000	−0.013	0.463	0.001	−0.177	−0.531
WW21AxiomaMH.Personal Products	0.000	−0.006	0.481	0.000	−0.329	−0.986

(continued)

Table 7.13 (continued)

	Contribution	Avg. Wtd. Exp.	HR	Risk	IR	T-Stat
WW21AxiomaMH.Health Care Equipment & Supplies	0.000	0.003	0.472	0.001	−0.126	−0.377
WW21AxiomaMH.Energy Equipment & Services	0.000	−0.009	0.481	0.002	−0.083	−0.250
WW21AxiomaMH.Gas Utilities	0.000	−0.001	0.481	0.000	−0.880	−2.641
WW21AxiomaMH.Distributors	0.000	0.010	0.435	0.001	−0.265	−0.795
WW21AxiomaMH.Household Products	0.000	−0.017	0.565	0.002	−0.160	−0.480
WW21AxiomaMH.Life Sciences Tools & Services	0.000	0.000	0.241	0.001	−0.377	−1.131
WW21AxiomaMH.Multi-Utilities	0.000	−0.006	0.556	0.001	−0.416	−1.247
WW21AxiomaMH.Automobiles	0.000	−0.004	0.537	0.001	−0.280	−0.839
WW21AxiomaMH.Diversified Financial Services	0.000	0.015	0.500	0.001	−0.261	−0.783
WW21AxiomaMH.Commercial Services & Supplies	0.000	0.009	0.528	0.001	−0.437	−1.311
WW21AxiomaMH.IT Services	0.000	−0.007	0.509	0.001	−0.288	−0.865
WW21AxiomaMH.Insurance	0.000	0.032	0.491	0.003	−0.128	−0.383
WW21AxiomaMH.Chemicals	0.000	0.005	0.463	0.001	−0.341	−1.023
WW21AxiomaMH.Oil, Gas & Consumable Fuels	0.000	0.000	0.407	0.003	−0.123	−0.370
WW21AxiomaMH.Capital Markets	0.000	−0.005	0.463	0.001	−0.463	−1.389
WW21AxiomaMH.Road & Rail	0.000	−0.010	0.444	0.001	−0.505	−1.514
WW21AxiomaMH.Airlines	−0.001	0.031	0.444	0.005	−0.120	−0.361
WW21AxiomaMH.Specialty Retail	−0.001	−0.004	0.463	0.002	−0.368	−1.104
WW21AxiomaMH.Real Estate Management & Development	−0.001	0.010	0.519	0.001	−0.708	−2.125
WW21AxiomaMH.Beverages	−0.001	−0.022	0.407	0.002	−0.566	−1.697
WW21AxiomaMH.Biotechnology	−0.001	0.033	0.509	0.005	−0.206	−0.619
WW21AxiomaMH.Machinery	−0.001	−0.009	0.343	0.001	−1.589	−4.767
WW21AxiomaMH.Auto Components	−0.001	0.002	0.509	0.001	−1.237	−3.712
WW21AxiomaMH.Marine	−0.001	0.016	0.426	0.004	−0.276	−0.828
WW21AxiomaMH.Paper & Forest Products	−0.001	0.006	0.556	0.001	−0.799	−2.396
WW21AxiomaMH.Food & Staples Retailing	−0.001	−0.025	0.500	0.002	−0.596	−1.789
WW21AxiomaMH.Electric Utilities	−0.001	0.005	0.463	0.001	−1.008	−3.025
WW21AxiomaMH.Tobacco	−0.001	−0.011	0.407	0.001	−1.051	−3.153

universe goes from 466,550 to approximately 7,500 stocks. We will refer to this universe as the GLER2012 universe. If one runs STE optimization with (1) No Risk Constraints; (2) 8% monthly turnover; (3) 150 basis points of transactions costs, each way; (4) a threshold position weight of 35 basis points; (5) and a maximum security weight of 4%; (5) long-only portfolio such that the minimum

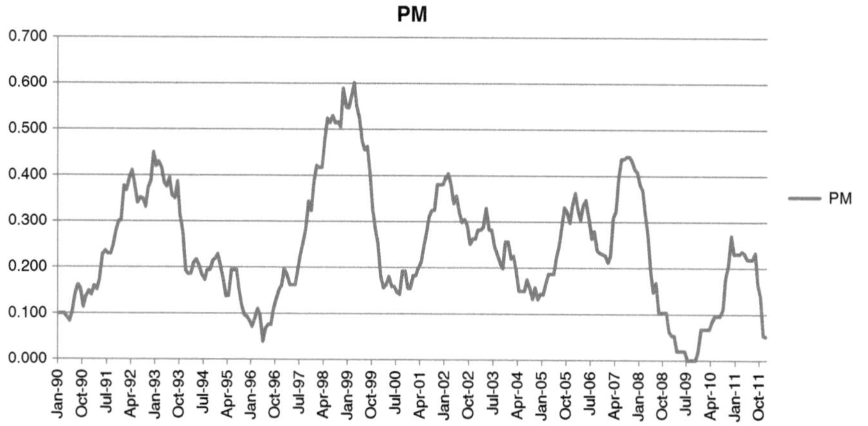

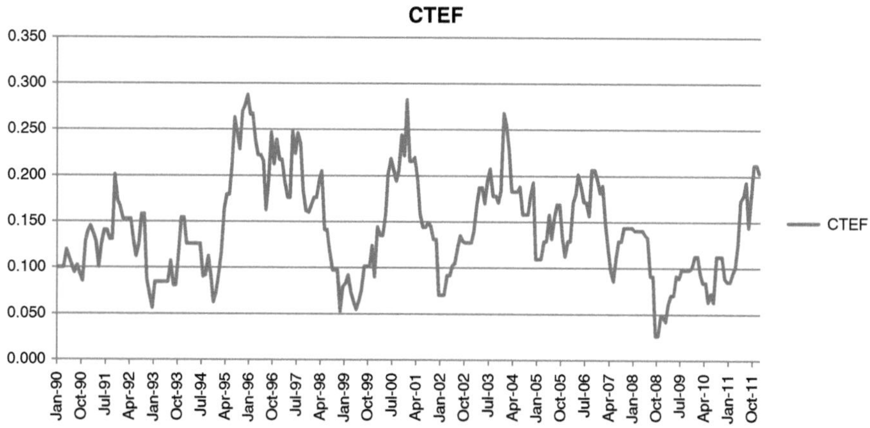

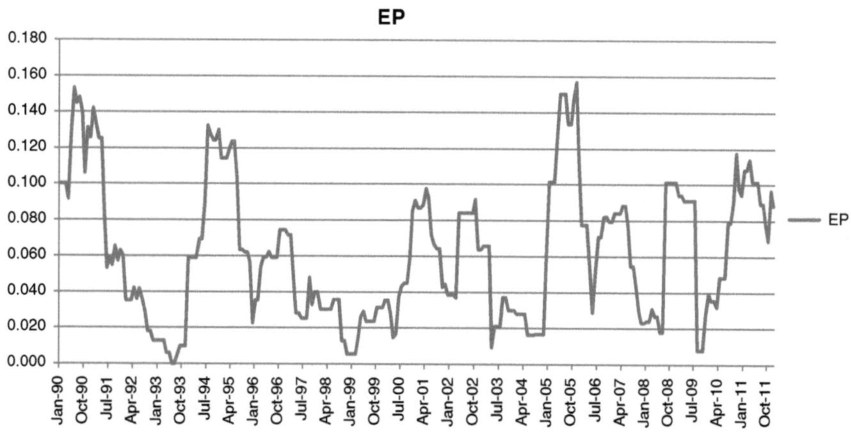

Fig. 7.4 (continued)

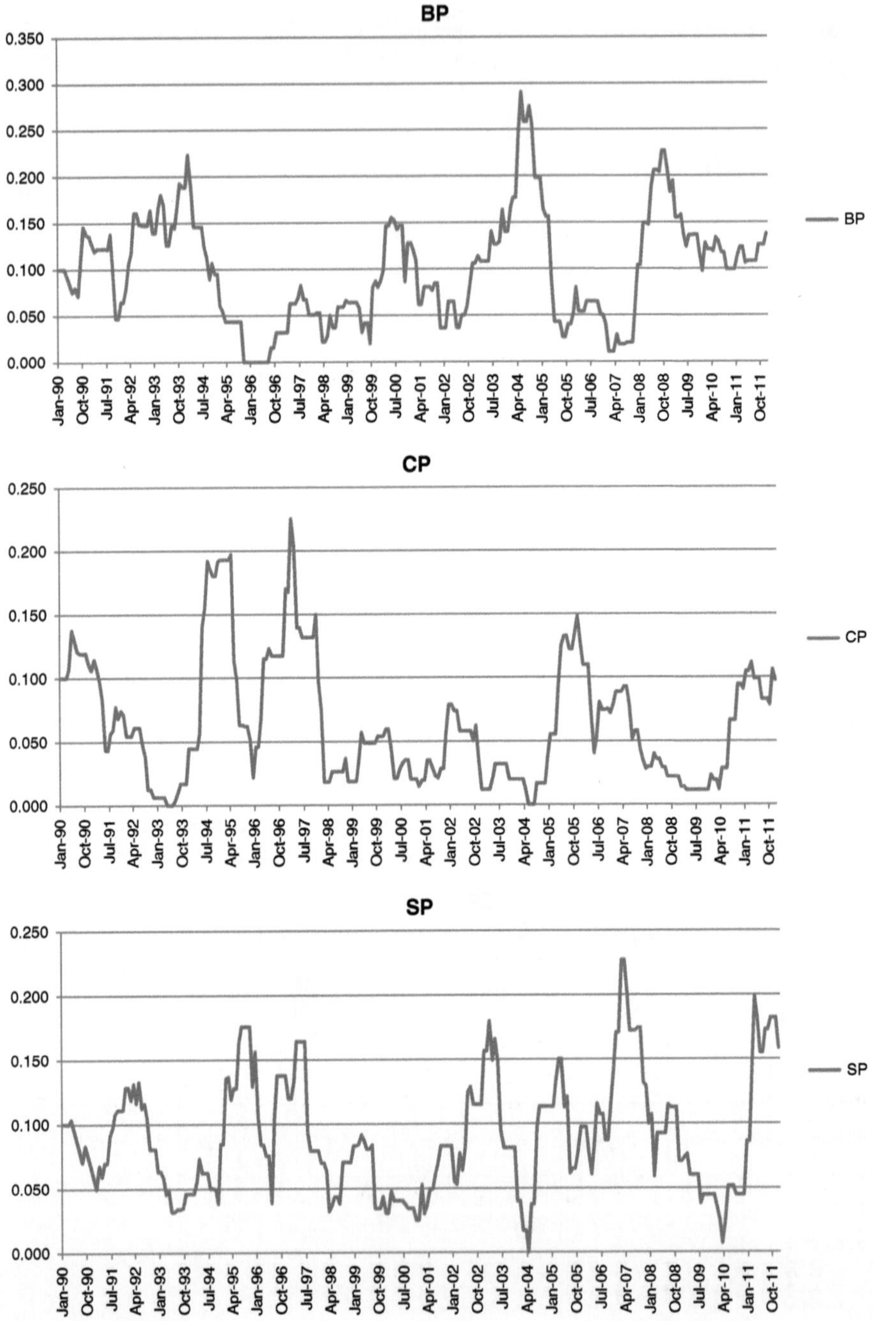

Fig. 7.4 (continued)

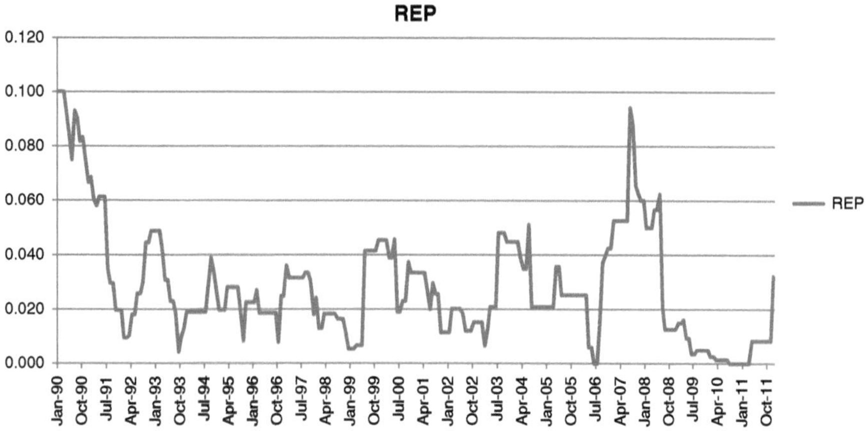

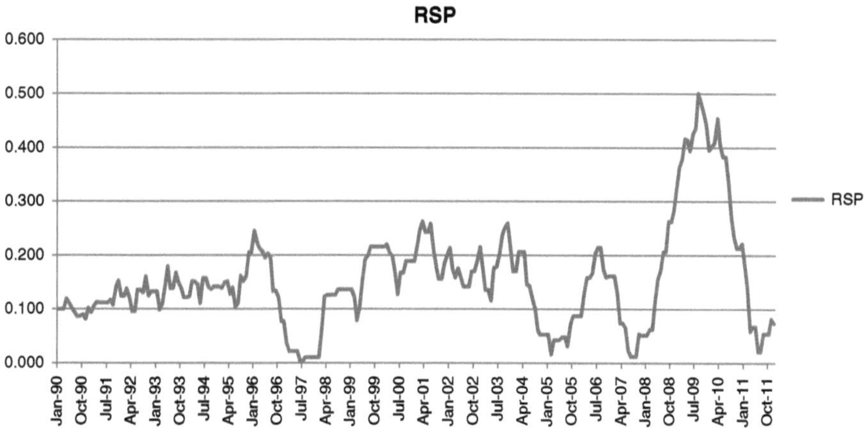

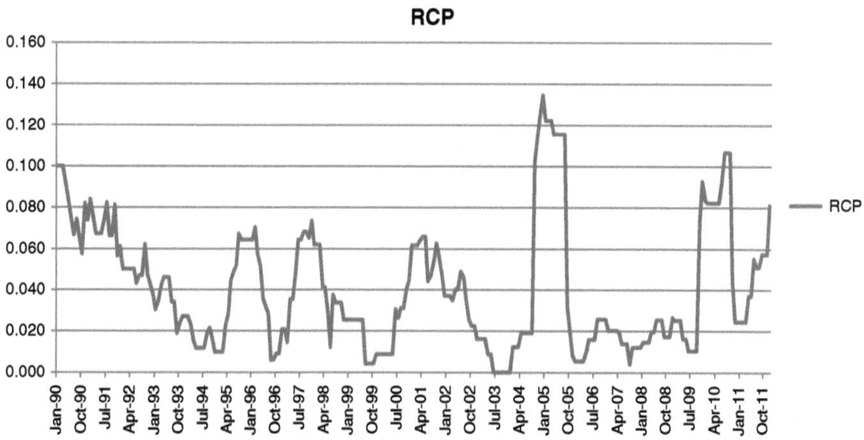

Fig. 7.4 (continued)

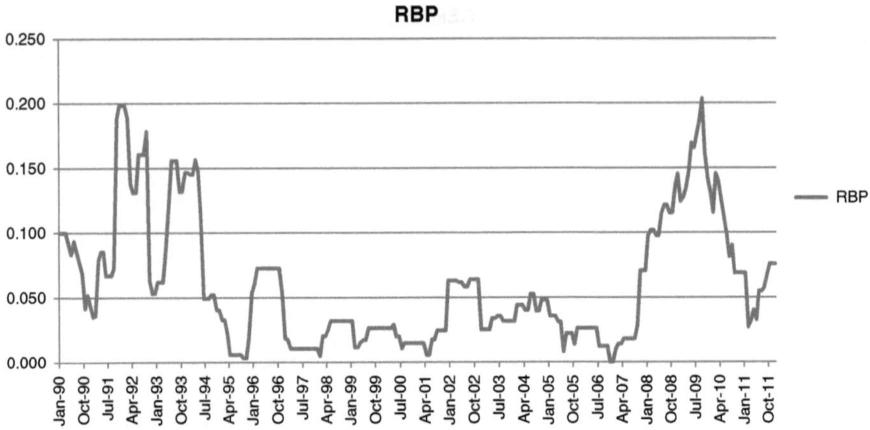

Fig. 7.4 Relative global model component weights, 1990–2011

weight is 0; and one uses lambda values of 500 and 200, then one produces portfolios producing higher Geometric Means, Sharpe Ratios, and Information Ratios than the universe benchmark (see Table 7.12). The Axioma attribution reveals statistical significant active return of 10.7% annually (see Table 7.13). The FactSet GLER regression weights are graphed in Fig. 7.4. In the FactSet universe, CTEF and PM amount to only 38% of the GLER model weights. PM has the largest weight, at about 24%.

There should be three results from the USER data analysis. An asset manager should set tracking errors at 8% to maximize the Geometric Mean, Sharpe Ratio, and Information Ratio; higher lambdas are preferred to lower lambdas (use at least an APT lambda of 100); and the Alpha Alignment Factor is most appropriate.

Conclusions

We addressed several issues in portfolio construction and management with the Guerard et al. (2012) USER data. First, we report that the Markowitz mean–variance (MV) optimization technique dominates the Enhanced Index-Tracking optimization technique at most security weight ranges. Second, we report that the Systematic Tracking Error optimization technique reported Wormald and van der Merwe (2011) is very effective in USA and global markets. Finally, we report that the Saxena and Stubbs (2012) Axioma Alpha Alignment Factor (AAF) is appropriate for USER and GLER Data and that the Axioma Statistical Risk Model dominates the Axioma Fundamental Model. The Markowitz approach to portfolio construction and management is 60 years old and remains an integral tool of investment research. Earnings forecasts play a very important role in identifying mispriced securities.

References

APT Analytics Guide 2011. SunGard APT, London www.sungard.com/apt/learnmore/

Bertsimas, D., C. Darnall, and R. Soucy. 1999. "Portfolio Construction Through Mixed-Integer Programming at Grantham, Mayo, Van Otterloo, and Company". *Interfaces* 29, 49–66.

Bertsimas, D., G.J. Lauprete, and A. Samarov. 2004. "Shortfall as a Risk Measure: Properties, Optimzation and Applications". *Journal of Economic Dynamics and Control* 28, 1353–1381.

Black, F., M.C. Jensen, and M. Scholes.1972. "The Capital Asset Pricing Model: Some Empirical Tests," in *Studies in the Theory of Capital Markets,* edited by M. Jensen. New York: Praeger.

Blin, J.-M., S. Bender, and J.B.Guerard Jr. 1997. "Earnings Forecasts, Revisions and Momentum in the Estimation of Efficient Market-Neutral Japanese and U.S. Portfolios." In *Research in Finance* 15, edited by A. Chen.

Bloch, M., J.B. Guerard Jr., H.M. Markowitz, P. Todd, and G. L. Xu. 1993. "A Comparison of Some Aspects of the U.S. and Japanese Equity Markets." *Japan & the World Economy* 5, 3-26.

Ceria, S., A. Saxena, and R.A. Stubbs. 2012. "Factor Alignment Problems and Quantitative Portfolio Management". *Journal of Portfolio Management* 28,

Chopra, V. K and W.T. Ziemba. 1993 "The Effects of Errors in Means, Variances and Covariances on Optimal Portfolio Choice", *Journal of Portfolio Management* 19, 2.

Connor, G. and R.A. Korajczyk. 1995. "The Arbitrage Pricing Theory and Multifactor Models of Asset Returns", Chapter 4 of *Finance, Handbooks in Operations Research and Management Science, Volume 9*, ed R. Jarrow, V. Maksimovic, and W. Ziemba. Amsterdam: North Holland, 1995

Connor, G. and R.A. Korajczyk 1993. "A Test for the Number of Factors in an Approximate Factor Model." *Journal of Finance* 48, 1263-1291.

Connor, G. and R.A. Korajczyk. 1988. "Risk and Return in an Equilibrium APT: Application of a New Test Methodology." *Journal of Financial Economics* 21, 255-289.

DeMiguel, V., L. Garlappi, F. J. Nogales and R. Uppal (2008), "A Generalized Approach to Portfolio Optimization: Improving Performance by Constraining Portfolio Norms", *Management Science* 55 No.5, 798-812.

Dharan, B.G.. 1997. Magnitude of goodwill in U.S. corporate financial statements. In Stephen A. Zeff and Bala G. Dharan, Editors, Readings and Notes on Financial Accounting. New York: McGraw-Hill, Inc.

Dhrymes, Irwin Friend and N. Biilent Gultekin 1984. "An *Empirical Examination of the Implications of Arbitrage Pricing Theory" Journal of Banking and Finance* 9, 73–99. March 1985.

Dhrymes, P.J., I. Friend, M.N. Gultekin, and N.B. Gultekin. 1985. "New Tests of the APT and Their Implications," *Journal of Finance* 40, 659–673.

Dhrymes, P.J., I. Friend, and N.B. Gultekin. 1984. "A Critical Reexamination of the Empirical Evidence on the Arbitrage Pricing Theory, *Journal of Finance* 39, 323–345.

Disatnik, D. and S. Benninga (2007) "Shrinking the Covariance Matrix—Simpler is Better", *Journal of Portfolio Management* 33, no 4, 56-63.

Elton, E.J., M.J. Gruber, S.J. Brown, and W.N. Goetzman. 2007. *Modern Portfolio Theory and Investment Analysis.* John Wiley & Sons, Inc., Seventh Edition.

Elton, E.J., M.J. Gruber, and M.N. Gültekin. 1981. "Expectations and Share Prices." *Management Science* 27, 975-987.

Fabozzi, F., F. Gupta &. Harry Markowitz. 2002. "The Legacy of Modern Portfolio Theory".

Fabozzi, F., F. Jones & R. Vardharaj. 2002. "Multi-Factor Equity Risk Models", in Fabozzi & Markovitz eds *The Theory and Practise of Investment Management.* New York: John Wiley & Sons.

Fabozzi,F., F. Gupta & Harry Markowitz. 2002a. "The Legacy of Modern Portfolio Theory" *Journal of Investing*, 7-22.

Fama, E.F. and K.R. French. 1992. "Cross-Sectional Variation in Expected Stock Returns". *Journal of Finance* 47, 427-465.

Fama, E.F. and K.R. French. 1995. "Size and the Book-to-Market Factors in Earnings and Returns". *Journal of Finance* 50:131–155.

Fama, E.F., and K.R. French. 1996. "Multifactor Explanations of Asset Pricing Anomalies". *Journal of Finance* 51, 55-84.

Fama,E.F. and K.R. French. 2008. "Dissecting Anomalies". *Journal of Finance* 63, 1653–1678.

Gültekin, M. N. and N. B. Gültekin. 1987. "Stock Return Anomalies and Tests of the APT." *Journal of Finance* 42, 1213–1224.

Grinhold, R. and R. Kahn. 2000. *Active Portfolio Management* New York: McGraw-Hill/Irwin, Second Edition.

Guerard, J.B. 1997. "Is There a Cost to be Socially Responsible in Investing?" *Journal of Forecasting* 16, 475–490.

Guerard, J.B. 2010. *The Handbook of Portfolio Construction: Contemporary Applications of Markowitz Techniques.* New York: Springer. Chapter 3.

Guerard, J.B. 2012. Global Earnings Forecast Efficiency (2012), *Research in Finance* 28, 19–47.

Guerard, J.B. Jr., M Takano, and Y. Yamane. 1993. "The Development of Efficient Portfolios in Japan with Particular Emphasis on Sales and Earnings Forecasting." *Annals of Operations Research* 45, 91-108.

Guerard, J.B., Jr. 2006. "Quantitative Stock Selection in Japan and the US: Some Past and Current Issues." *Journal of Investing,* 15.

Guerard, J.B., Jr. 2010. Global Earnings Forecast Efficiency (2010), *McKinley Capital*, presented at 30th International Symposium on Forecasting, June 2010.

Guerard, J.B., Jr., E. Krauklis, and M. Kumar. 2012. "Further Analysis of Efficient Portfolios with the USER Model", *Journal of Investing* 21, 81–88.

Guerard, J.B., Jr., M.N. Gultekin, and G. Xu. 2012. "Investing with Momentum: The Past, Present, and Future", *Journal of Investing* 21, 68-80.

Guerard, J.B., S. C. Chettiappan, and G-L, Xu. 2010. "Stock-Selection Modeling and Data Mining Corrections: Long-Only versus 130/30 Models". *The Handbook of Portfolio Construction: Contemporary Applications of Markowitz Techniques.* New York: Springer.

Jagannathan, R., and T. Ma. 2003. "Risk Reduction in Large Portfolios: Why Imposing the Wrong Constraints Helps", *Journal of Finance* 58, 1651-1684.

King, B.F. 1966. "Market and Industry Factors in Stock Price Behavior." *Journal of Business* 39, 139–191.

Kolbert, F. and L. Wormald. 2010. Chapter 1 in S. Satchell, Ed. *Optimizing Optimization : The Next Generation of Optimization Applications and Theory*, Amsterdam: Elsevier Quantitative Finance Series.

Latane, H.A. 1959."Criteria for Choice Among Risky Ventures". *Journal of Political Economy* 67, 144–155.

Latane, H.A., D.L. Tuttle, and C.P. Jones. 1975. *Security Analysis and Portfolio Management.* New York: The Roland Press. 2nd Edition.

Ledoit, O. and M. Wolf. 2003."Improved Estimation of the Covariance Matrix of Stock Returns with an Application to Portfolio Selection", *Journal of Empirical Finance*, Volume 10, Issue 5, Dec 2003, pages 603-621

Ledoit, O. and M. Wolf. 2004. "Honey, I Shrunk the Sample Covariance Matrix", *Journal of Portfolio Management* 31, Number 1, Fall 2004.

Lee, J.H. and D. Stefek. 2008. "Do Risk Factors Eat Alphas?" *Journal of Portfolio Management* 34, 4, 12–25.

Levy, H. 1999. Introduction to Investments. Cincinnati: South-Western College Publishing. Second Edition.

Lintner, J. 1965. "Security Prices, Risk, and the Maximum Gain from Diversification." *Journal of Finance* 30.

Lintner, J. 1965. "The Valuation of Risk Assets on the Selection of Risky Investments in Stock Portfolios and Capital Investments." *The Review of Economics and Statistics,* 13–37.

Markowitz, H. M. 2012."Topics in Applied Investment Management: From a Bayesian Viewpoint, *Journal of Investing* 21, 7–14.

Markowitz, H.M. 1952. "Portfolio Selection." *Journal of Finance* 7, 77–91.

Markowitz, H.M. 1959. *Portfolio Selection: Efficient Diversification of Investment.* Cowles Foundation Monograph No. 16. New York, John Wiley &. Sons.

Markowitz, H.M. 1976. "Investment in the Long Run: New Evidence for an Old Rule," *Journal of Finance* 31, 1273–1286.

Markowitz, H.M. 2000. *Mean-Variance Analysis in Portfolio Choice and Capital Markets.* New? Hope, PA: Frank J. Fabozzi Associates.

Markowitz, H.M. *Mean-Variance Analysis in Portfolio Choice and Capital Markets.* Oxford: Basil Blackwell, 1987.

Markowitz, H.M., and G. Xu. 1994. "Data Mining Corrections." *Journal of Portfolio Management* 21, 60–69.

Mulvey, J. W. Chang Kim, & M. Bilgili. 2010. "Linking Momentum Strategies with Single-Portfolio Models", in J. Guerard, Ed. *Handbook of Portfolio Construction: Contemporary Applications of Markowitz Techniques*, New York: Springer.

Onnela, J.P., A Chakraborti, K Kaski, J Kertész and A Kanto.2003. "Dynamics of Market Correlations: Taxonomy and portfolio Analysis", *Physical Review E* 68, 056110.

Pastor, L. and R. F. Stambaugh. 2003. "Liquidity Risk and Expected Stock Returns." Journal of Political Economy, 111, 642–684.

Rachev, S. and S. Mittnik. 2000. *Stable Paretian Models in Finance.* New York: Wiley.

Ramnath, S., S. Rock, and P. Shane. 2008. "The Financial Forecasting Literature: A Taxonomy with Suggestions for Further Research." *International Journal of Forecasting*, 24, 34-75.

Robinson, R. H. Greuning, E. Henry E., and M. Broihahn. 2009. In International Financial Statement Analysis. CFA Institute Investment Series.

Roll, R. 1969. "Bias in Fitting the Sharpe Model to Time series Data", *Journal of Financial and Quantitative Analysis* 4, 271–289.

Roll, R. 1977. "A Critique of the Asset Pricing Theory's Tests". *Journal of Financial Economics* 4, 129–176.

Rosenberg, B. 1974. "Extra-Market Components of Covariance in Security Returns." *Journal of Financial and Quantitative Analysis* 9, 263-274.

Rosenberg, B. and V. Marathe, 1979. "Tests of Capital Asset Pricing Hypotheses." In *Research in Finance* 1, 1979, edited by H. Levy.

Ross, S.A. 1976. "The Arbitrage Theory of Capital Asset Pricing." *Journal of Economic Theory* 13, 341–360.

Ross, S.A. and R. Roll. 1980. "An Empirical Investigation of the Arbitrage Pricing Theory." *Journal of Finance* 35, 1071–1103.

Rudd, A., and H. K.Clasing 1982. *Modern Portfolio Theory, The Principles of Investment Management.* Homewood, IL: Dow-Jones Irwin.

Saxena, A. and R. A. Stubbs. Pushing Frontiers (literally) using Alpha Alignment Factor. Technical report, Axioma, Inc. Research Report #022, February.

Saxena, A. and R.A. Stubbs 2012. "An Empirical Case Study of Factor Alignment using the USER Model", *Journal of Investing* 21, 25–44.

Saxena, A. and R.A. Stubbs. 2010. Alpha Alignment Factor: A Solution to the Underestimation of Risk for Optimized Active Portfolios. Technical Report. Axioma, Inc. Research Report #015, February.

Saxena, A. and Robert A. Stubbs. 2012. "Synergistic Advantages of Custom Risk Models in Portfolio Construction: Risk Decomposition and Performance Attribution." Forthcoming.

Saxena, A., C. Martin, and R. A. Stubbs. 2011. "Aligning Alpha and Risk Factors: A Panacea to Factor Alignment Problems." Axioma Advisor. Research Focus.

Sharpe, W.F. 1963. "A Simplified Model for Portfolio Analysis." *Management Science* 9, 277–293.

Sharpe, W.F. 1964. "Capital Asset Prices: A Theory of Market Equilibrium under Conditions of Risk." *Journal of Finance* 19, 425–442.

Siegel, L. B. 2003. U.S. equity style indexes. In Benchmarks and Investment Management, chapter 8. The CFA Institute Research Foundation.

Solnik, B. 1974. "Why Not Diversify Internationally Rather than Domestically". *Financial Analysts Journal* 38, 48–54.

Solnik, B. 2000. *International Investments,* Reading, Mass: Addison Wesley Longman, Inc., 4th? Edition.

Stone, B. K. and J.B. Guerard, Jr. 2010. Methodologies for Isolating and Assessing the Portfolio Performance Potential of Stock Return Forecast Models with an Illustration, in J.B. Guerard, Jr., editor, *The Handbook of Portfolio Construction: Contemporary Applications of Markowitz Techniques.* New York: Springer.

Stone, B.K. 1970. *Risk, Return, and Equilibrium: A General Single-Period Theory of Asset Selection and Capital Market Equilibrium.* Cambridge, MA: MIT Press.

Vander Weide, J. H. 2010. "Principles for Lifetime Portfolio Selection: Lessons from Portfolio Theory", in J. Guerard, Ed. *Handbook of Portfolio Construction: Contemporary Applications of Markowitz Techniques.* New York: Springer.

Wormald, L. and M. van der Merwe. 2012. "Constrained Optimization for portfolio Construction" *Journal of Investing* 21, 44–59.

Chapter 8
Forecasting World Stock Returns and Improved Asset Allocation

In this chapter, we address the issue of market-timing. This chapter was written on a long and cold weekend in Anchorage when one of our salespersons called me and asked whether one could develop a model to identify periods when equities might be under-valued. The author was a senior in college in 1974 and was familiar with periods that could be buying opportunities.

There is little evidence in the literature on whether predictability of stock returns leads to improved asset allocation and performance (Handa and Tiwari 2006). Handa and Tiwari (2006) found fixed results for forecasting 1-month-ahead results in the USA for 1954–2002 period; the past-returns model worked well from 1974 to 1988 and poorly from 1959 to 1973 and 1989 to 2002. There are mixed academic results for many financial tests. In this report, we show that it is possible to improve performance of a naïve "60/40" model of equity and debt to a "60/40" model with

J.B. Guerard, Jr., *Introduction to Financial Forecasting in Investment Analysis*,
DOI 10.1007/978-1-4614-5239-3_8,

Global Timing (GT). We create a Global Timing signal based on the 12-month moving average of the differential between the LIBOR rate and the All Country World (ACW) index. If the predicted return signal, the differential of the 12-month average returns on the ACW, exceeds LIBOR by a statistically significant difference (one standard deviation), then a "buy" signal is created. If the predicted return signal is less than −1.645, one standard deviation, then a "sell" decision is made. A neutral position exists in the signal and no change is made.[1]

As with Handa and Tiwari (2006), we restrict our investment choices to a relatively riskless asset, LIBOR, or an investment in ACWG securities. We test the model on ACW index and implement on the ACW or ACWG indexes. We are a growth manager and use the constituent securities in the ACWG index. The asset allocation benchmark is a "60/40" portfolio invested in 60 percent in a passive

---

[1] A similar signal was developed to investigate the relationship between Euro LEI and the GEM2 factor returns. For instance, suppose that a rise in the LEI one month could be associated with a rise in a GEM2 factor return three months later. An investor might then profit by taking a long position in the factor whenever the three-month lagged LEI were positive. One can use the Euro area Leading Economic indicator, LEI, series published by The Conference Board (TCB). Let $LEI(t)$ be the LEI level at the end of month t. Generally, these values are published with a 1- or 2-month lag. The "return" to the LEI over month t is then given by

$$L_t = \frac{LEI(t) - LEI(t-1)}{LEI(t-1)}. \tag{8.1}$$

The lagged correlation between the GEM2 factor return and the LEI return is

$$\rho_k^m = corr\left(f_{kt}^P, L_{t-m}\right), \tag{8.2}$$

where f_{kt}^P is the pure return to factor k over period t, and m is the number of lags in months.

Optimal portfolios are created using the MSCI Barra GEM2 risk model, the premier institutional asset manager portfolio management, and control system. The GEM2 model, described in Menchero et al. (2010), estimates a multifactor risk model composed of eight factors: the world, value, growth, momentum, liquidity, size, size nonlinearity, and leverage. The GEM2 Model is the global equivalent of the USE3 model used in Chap. 6. The Barra model allows the asset manager to specifically target desired portfolio exposures to accommodate client needs and expectations, such as having an exposure to momentum and not necessarily having other exposures. Simple factor portfolios have unit exposure to the particular factor, and nonzero exposure to other factors. Pure factor portfolios have unit exposure to the particular factor, and zero exposure to all other factors. Optimal factor portfolios have the minimum risk portfolio with unit exposure to the factor. Menchero et al. (2012) reported the strongest positive correlation that suggests a positive relationship between changes in the LEI and corresponding changes in Momentum six months later. A momentum-timing signal is created in which if an increase in 6-month average change in LEI exceeds 1.50 standard deviations, then one becomes aggressive with respective to momentum. One sells momentum if the 6-month average change in momentum is less than 1.50 standard deviations. We also present the cumulative performance of the pure Momentum factor, as well as the Euro LEI series. The momentum timing returns have been scaled to have the same realized volatility as the pure momentum factor over the 13-year period. Menchero, Morozov, and Shepard (2012) reported that the momentum timing strategy greatly outperforms the pure momentum strategy over this sample period, with the former climbing more than 60 %, compared to only 20 % return for the pure factor, primarily identifying the 2009 plunge.

Table 8.1 Attribution report of the TAA signal portfolios, 1/2002–10/2011

Annualized contributions to total return				
Source of return	Contribution (% return)	Risk (% std. dev.)	Info ratio	T-stat
1. Risk free	1.86			
2. Total benchmark	4.67	17.46		
3. Currency selection	3.67	3.52	1.09	3.40
4. Cash-equity policy	0.00	0.00		
5. Risk indices	6.16	4.36	1.23	3.86
6. Industries	−0.38	2.56	−0.14	−0.44
7. Countries	0.97	5.07	0.19	0.61
8. World equity	0.00	0.00		
9. Asset selection	1.16	3.22	0.41	1.29
10. Active equity [5 + 6 + 7 + 8 + 9]	7.91	7.60	0.96	3.02
11. Trading				
12. Transaction cost	−4.25			
13. Total active [3 + 4 + 10 + 11 + 12]	7.63	8.22	0.93	2.91
14. Total managed [2 + 13]	12.29	19.77		

Table 8.2 Strategy summary, January 2002–October 2011

Strategy	Cumulative Wealth ratio	Mean Monthly return	Sharpe ratio
"60/40"	3.804	1.185	1.143
ACWG index	1.566	0.506	0.88
"60/40" GT	5.826	1.567	1.367

basket of ACW securities. If the Tactical Asset Allocation (TAA) signal exceeds 1.645, then we buy. If the TAA signal is less than −1.645, then we sell. How can we implement such a strategy in a long-only investment portfolio? As with the McKinley Capital Management (MCM) "Global Alpha-Engineering a Dynamic Momentum" strategy, we may vary the portfolio lambda, the measure of the return–risk preference of the asset manager. If the TAA signal exceeds 1.645, then we implement a portfolio lambda of 200, leading to an aggressive return-to-risk portfolio. If the TAA signal is less than −1.645, then we implement a lambda of 10, indicating a relatively passive return-to-risk portfolio. The reader is reminded of the APT lambda – estimated trade offcurve in Chapter 7. Friends do not let friends invest with a lambda less than 100. If the TAA signal is neutral, then we use a lambda of 75. We use MQ, a quantitative-based strategy described in the MCM "Global Alpha" research report, as the portfolio expected return.

An investor can purchase instruments or ETFs to produce a "60/40" return for the February 1997–October 2011 period. We ran the simulations from January 1997 to October 2011, varying the portfolio returns using monthly signals and targeting the All Country World Growth (ACWG) Index. We measure the performance of the simulations from January 2002 to October 2011, the period of the Global (GEM2) Model. The TAA signals portfolio produces statistically significant total active returns, see Table 8.1.

Had an investor invested in a "60/40" strategy, the mean monthly return of 1.185 percent for January 2002–October 2011 exceeds the ACWG Index return of 0.506 for the corresponding period. The "60/40"GT strategy produces a monthly return of 1.567 (including transactions costs of 150 basis points each way). The TAA signals portfolio outperforms the market and the "60/40" strategy in producing higher Sharpe Ratios. Thus, the TAA portfolios produce higher returns for a given level of risk than the "60/40" strategy and the ACWG index (Table 8.2).

CHART 1: TAA Analysis, 1/2002 -10/2011

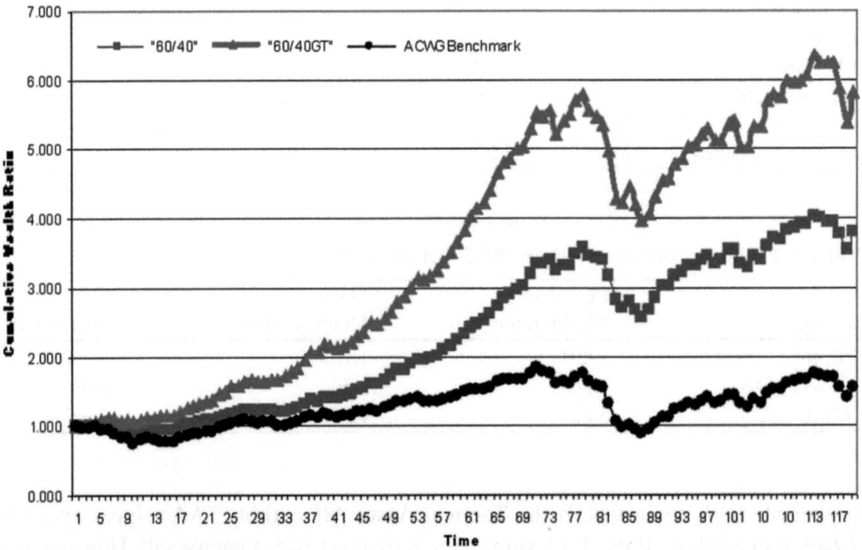

Summary and Conclusions

Stock return expectations can be used to vary the aggressiveness of equity portfolios that can lead to Tactical Asset Allocation decisions that can outperform a naïve "60/40" strategy.

References

Guidolin, M. and A. Timmermann. 2004. "Economic Implications of Bull and Bear Regimes in UK Stock and Bond Returns". Unpublished Working Paper.

Handa, P. and A. Tiwari. 2006. "Does Stock Return Predictability Imply Improved Asset Allocation and Performance? Evidence from the U.S. Stock Markey (1965–2002). Journal of Business 79, 2423–2468.

Markowitz, H.M. 1959. *Portfolio Selection: Efficient Diversification of Investment*. Cowles Foundation Monograph No.16. New York, John Wiley & Sons.
McKinley Capital Management, LLC. 2010. "Global Alpha-Engineering a Dynamic Exposure".
Menchero, J., A. Morozov, and J.B. Guerard, Jr. 2012. "Capturing the Eqity Risk Premia", Chapter 25 in C.F. Lee, J. Finnerty, J. Lee, and D. Wort. *Security Analysis, Portfolio Management, and Financial Derivatives*. Singapore: World Scientific.
Menchero, J., Morozov, A., and P. Shepard, 2010. "Global Equity Risk Modeling," in J. Guerard, Ed., *The Handbook of Portfolio Construction: Contemporary Applications of Markowitz Techniques*, Springer.
MSCI Barra Fundamental Data Methodology Handbook, 2008, http://www.msci.com/methodology/meth_docs/MSCI_Sep08_Fundamental_Data.pdf
The Conference Board LEI for the Euro Area, 2001. Business Cycle Indicators Handbook.

Chapter 9
Summary and Conclusions

The forecasting of earnings per share, eps, is a most important input to an investment strategy. The stock selection model estimated in Chapters 4 and 6, and used as an expected return input to the Markowitz portfolio optimation problem in Chapter 7 is highly effective at identifying mis-priced stocks and can be used to produce portfolios with highly statistically significant asset selection. There is a tremendous literature regarding forecasting of corporate eps and whether the forecasts are more accurate than a random walk or a random walk with drift. Much of the literature can be summarized as follows: (1) analysts' forecasts are not statistically different from a random walk with drift model; that is, analysts' forecasts can be approximated with a first-order exponential smoothing model forecast; (2) analysts' forecasts are biased; analysts' forecasts are optimistic; (3) analysts' forecast revisions and the direction of their revisions are more highly correlated with stock returns than earnings forecasts themselves; (4) earnings forecasts are highly statistically significant in forecasting total stock returns; (5) earnings forecasts, revisions, and direction of revisions can be combined with fundamental data, such as earnings, book value, cash flow, sales, these variables relative to their histories, and price momentum strategies to identify mispriced stocks; (6) smaller capitalized stocks are more mispriced than larger capitalized stocks; and (7) international and global stocks are more mispriced than the US stocks.

We introduced the reader to regression models and various estimation procedures. We have illustrated regression estimations by modeling consumption functions and the relationship between real GDP and The Conference Board Leading Economic Indicators (LEI). We estimated regressions using EViews, SAS, and automatic modeling in Oxmetrics. There are many advantages with the various regression software with regard to ease of use, outlier estimations, collinearity diagnostics, and automatic modeling procedures.

We introduced reader to the time series work of Professors Box and Jenkins and examined the predictive information in The Conference Board LEI for the USA, the UK, Japan, and France. We find that The Conference Board LEI and FIBER short-term LEI are statistically significant in modeling the respective real GDP changes during the 1970–2000 period. One rejects the null hypothesis of no association

J.B. Guerard, Jr., *Introduction to Financial Forecasting in Investment Analysis*,
DOI 10.1007/978-1-4614-5239-3_9, © Springer Science+Business Media New York 2013

between changes in the LEI and changes in real GDP in the USA, and the G7 nations. If one uses a rolling 32 quarter estimation period and a one-period-ahead forecasting root mean square error calculation, the LEI forecasting errors are not significantly lower than the univariate ARIMA model forecasts.

We used two case studies to illustrate the effectiveness of regression modeling. Regression analysis offered marginal improvement in the case of combining GNP forecasts, but offered substantial improvement in identifying financial variables associated with security returns. We introduced the reader to a stock selection model that combined earnings forecasts, fundamental variables derived from balance sheet and income statement analysis, and price momentum variables. The regression-based United States Expected Returns (USER) Model was highly statistically significant in construction. Regression techniques addressing outliers and multicollinearity problems in the USER Model outperformed equally weighted strategies in stock selection modeling.

A case study of mergers was introduced so that the reader could examine Granger causality testing in detail. Mergers were modeled as a function of the LEI and stock prices.

The Barra Aegis system has been the industry standard for portfolio construction, management, and measurement for almost 40 years. We demonstrated the effectiveness of the Barra Aegis system to create investment management strategies to produce portfolios and attribute portfolio returns to the Barra multifactor risk model during the December 1979–2009 period. We find additional evidence to support the use of MSCI Bara multifactor models for portfolio construction and risk control. We report two results: (1) a composite model incorporating fundamental data, such as earnings, book value, cash flow, and sales, with analysts' earnings forecast revisions and price momentum variables to identify mispriced securities in the Guerard, Xu, and Gultelein (2012) USER Model. (2) the returns to a multifactor risk-controlled portfolio allow us to reject the null hypothesis that the results are due to data mining. We develop and estimate three levels of testing for stock selection and portfolio construction. The use of multifactor risk-controlled portfolio returns allows us to reject the null hypothesis that the results are due to data mining. The anomalies literature can be applied in real-world portfolio construction.

We addressed several additional issues in portfolio construction and management with the USER data. First, we report that the Markowitz Mean-Variance (MV) optimization technique dominates the Enhanced Index-Tracking optimization technique at most security weight ranges. Second, we report that the Systematic Tracking Error optimization technique reported by Wormald and van der Merwe (2012) is very effective in the US and Global markets. Finally, we report that the Saxena and Stubbs (2012) Axioma Alpha Alignment Factor (AAF) is appropriate for USER and global (GLER) Data and that the Axioma Statistical Risk Model dominates the Axioma Fundamental Model. The Markowitz (1959) approach to portfolio construction and management is sixty years old and remains an integral tool of investment research. Markowitz (2012) continues to believe that his last four Chapters of his *Portfolio Selection* are extremely relevant. The author agrees, however, one can fine-tune the Markowitz portfolio construction and management process and enhance its applicability to the current stock market environment. Earnings forecasts play a very important role in identifying mispriced securities.

Finally, stock return expectations can be used to vary the aggressiveness of equity portfolios that can lead to Tactical Asset Allocation decisions that can outperform a naïve "60/40" strategy.

In July 2011, McKinley Capital hosted a research conference in Anchorage, Alaska where Axioma, Sungard APT, Barra, and FactSet presented research into United States stock portfolio construction and management.

In July 2013, McKinley Capital expects to host a global portfolio construction and management research seminar. The relevant them is "Applied Investment Research". Forecasting of earnings is an extremely important and (statistically) significant part of stock selection modeling.

References

Guerard, J.B. Jr., G. Xu, and M.N. Gultekin. 2012. "Investing with Momentum: The Past, Present, and Future", *Journal of Investing* 21, 68–80,

Markowitz, H.M. 1959. Portfolio Selection: Efficient Diversification of Investments. New York: Wiley.

Markowitz, H.M. 2012. "Topics in Applied Investment Management: From a Bayesian View Point", *Journal of Investing* 21, 7–13.

Saxena, A. and R.A. Stubbs. 2012. "An Empirical Case Study of Factor Alignment Problems Using the USER Model", *Journal of Investing* 21, 25–43.

Wormald, L. and E. Van der Merwe. 2012. "Constrained Optimization for Portfolio Construction", *Journal of Investing* 21, 44–59.

Index

J.B. Guerard, Jr., *Introduction to Financial Forecasting in Investment Analysis*,
DOI 10.1007/978-1-4614-5239-3, © Springer Science+Business Media New York 2013

MIX
Papier aus verantwortungsvollen Quellen
Paper from responsible sources
FSC® C105338

If you have any concerns about our products,
you can contact us on
ProductSafety@springernature.com

In case Publisher is established outside the EU,
the EU authorized representative is:
Springer Nature Customer Service Center GmbH
Europaplatz 3, 69115 Heidelberg, Germany

Printed by Libri Plureos GmbH
in Hamburg, Germany